Essentials of Mechanical Stress Analysis

Essentials of Mechanical Stress Analysis, updated for the second edition, covers stress analysis from an interdisciplinary perspective. Discussing techniques and theories essential to analysing structures, the book covers both analytical and numerical approaches.

The second edition adds new topics and updates research to follow current advances in the field. New sections on material properties, composite materials and finite element analysis enable the reader to further establish the fundamental theory behind material behaviour and the causes of stress and strain. Also covering beams, plates, columns and elastic instability, the book discusses fatigue, life cycle, energy methods and MathCAD sample code.

As a clear and comprehensive guide to stress and structural analysis, this book is relevant to students and scholars in the fields of mechanical, aerospace and civil engineering, as well as materials science.

Essentials of Mechanical Stress Analysis

Second Edition

Amir Javidinejad

CRC Press
Taylor & Francis Group
Boca Raton London New York

CRC Press is an imprint of the
Taylor & Francis Group, an **informa** business

Second edition published 2023
by CRC Press
6000 Broken Sound Parkway NW, Suite 300, Boca Raton, FL 33487-2742

and by CRC Press
4 Park Square, Milton Park, Abingdon, Oxon, OX14 4RN

CRC Press is an imprint of Taylor & Francis Group, LLC

First edition published by CRC Press 2015

ISBN: 978-1-032-31755-7 (hbk)
ISBN: 978-1-032-31765-6 (pbk)
ISBN: 978-1-003-31121-8 (ebk)

DOI: 10.1201/9781003311218

Typeset in Times
by MPS Limited, Dehradun

Dedication

To my daughters Sara and Tina

Contents

Preface

This book is written to provide a text for associated topics in stress and structural analysis for higher level undergraduates, graduate students and practicing stress analyst in mechanical, aerospace, civil and materials engineering fields. It is written to provide the essential theories needed for problem solving and stress analysis on structural systems. The study of this book requires prior knowledge of undergraduate mechanics of material, engineering materials and engineering mathematics. Both theory and numerical examples are provided for better understanding of the concepts. The second edition of this book adds the S.I. unit examples to the English unit examples already provided in the book. Also, this second edition of the book has elaborated more on the topics that were deemed to be essential. The definitions of stress and strain were detailed in a more comprehensive manner. The topic of material selection in design is elaborated in this new edition. Stresses due to combined loading conditions and rotational loading are newly explored in this book. For beam designs in general, a guide is shown in tabular form to make an easily convenient method for the beam design process. At times, when finite element analysis (FEA) is used as a tool importance of the FEA results correlations to theory are elaborated. Also, determination of fastener stiffness and use as an input to the FEA model is explored for more accurate attachment point simulations.

In today's engineering world, more and more companies are requiring that the stress analysts carry out multidisciplinary topics of solid mechanics all at once. Taking that into consideration, this book is designed to cover a broad collection of topics in the stress analysis field that are essential for carrying out analysis of structures. It is unique because it gathers topics together that are otherwise normally presented as individual course topics. It covers both the analytical and numerical approaches to stress analysis. It covers isotropic, metallic and orthotropic, composite material analyses.

Chapter 1 covers the fundamentals of engineering materials, which is a prerequisite and requirement for understanding the concept of stress analysis. Chapter 2 elaborates on the basic concepts of stress and strain. It covers the relationship between stress and strain. It focuses on the state of stress, and it covers the principal stress calculations. Chapter 3 is designed to show the application of a polar coordinate system in analysis of stress and strain. It covers the concept of the stress field due to the line loading, stress concentration and stresses for pressure vessels. Chapter 4 introduces the different failure criteria and margins of safety calculations. Chapter 5 is written to illustrate beam analysis theory. It elaborates on the concepts of shear and moment diagrams, beam deflection, bending of beams, shear of beams and torsion of beams. Additionally, this chapter covers the curved-beam analysis theory. Chapter 6 is designed to cover the fundamentals of plate theory for stress and deflection analysis of circular and rectangular plates. Chapter 7 covers the topics of elastic instability and buckling of columns and plates. Chapter 8 covers the energy methods applicable for determining deflection and stresses of structural systems. Chapter 9 is written to illustrate the concept of fatigue and stress-to-life-cycle calculations. Chapter 10 introduces the numerical methods and finite element techniques used for stress analysis of structures. It covers the techniques for analysis of beams and rods. Also accurate Finite Element Modeling is discussed. Chapter 11 covers the stress analysis methods for composite material. It covers how lamina and laminate stress analysis are performed. Chapter 12 briefly illustrates fastener and joint connection analysis theory. Chapter 13 provides MathCAD computer worksheets that are developed for stress analysis simulations of the topics covered in chapters 1 through 12. This chapter is a very comprehensive collection of simulation codes that can be used for fast and reliable stress analysis of both metallic and composite sections.

Author

Amir Javidinejad earned a PhD in mechanical engineering at the University of Texas at Arlington and an MS in engineering mechanics at the Georgia Institute of Technology and holds a Certificate in Leadership Mastery from UCLA-Extension. He has extensive experience in structural/solid mechanics, finite element methods, machine design and various other stress analysis methods from aerospace, military and commercial industries, as well as from academia. His expertise and knowledge include space structures analysis, micro sensors analysis, rocket design analysis, helicopter structural repair analysis, airplane structures modifications, aircraft interior monument structures analysis, certification and qualification testing. He has also been involved in research in the areas of structural mechanics of isotropic, anisotropic and composite materials. Dr. Javidinejad is a Licensed Professional Mechanical Engineer in the state of California, License #38567, and in the state of Texas, License #141561. Dr. Javidinejad is a member of Pi Tau Sigma mechanical engineering honor society, a member of the American Society of Mechanical Engineers (ASME) and a member of the American Society of Engineering Education (ASEE). Dr. Javidinejad also is a lead structural analysis engineer at Boeing Company on a full-time basis and a part-time Lecturer of mechanical engineering at the California State Polytechnic University, Pomona.

1 Basics of Engineering Materials

1.1 INTRODUCTION

The stress analysis theory is based on the concepts derived from the material behavior. There are several relationships fundamental to the concepts of material behavior that need to be explored before studying the stress analysis theory. Terms such as Hooke's stress-strain behavior, modulus of elasticity of the material, Poisson's ratio effect, shear modulus, ultimate strength, shear strength, yield strength, stress concentration and coefficient of thermal expansion are usually used to define the material behavior.

1.2 THE FUNDAMENTAL TERMS

Under the application of any loading over a cross-section, the ratio of loading to the cross-section area is known as the stress. For instance, assuming a round bar constrained at one end and pulled by a force (F) on the other end, tensile loading over the cross-section area of the bar (A = Ao) is known as normal stress. Refer to Figure 1.1 for an illustration of this concept.

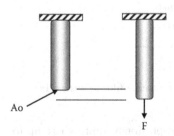

Ao

F

FIGURE 1.1 Bar object under axial tension (or compression).

Thus, the stress is defined as:

$$\sigma = \frac{F}{Ao} \text{ (unit of psi)} \tag{1.1}$$

Assuming the previous bar under applied loading, when it deforms and elongates, the deformation (δ) along the direction of loading per original length of the bar (L = Lo) is defined as strain (ε). Refer to Figure 1.2 for an illustration of this concept.

DOI: 10.1201/9781003311218-1

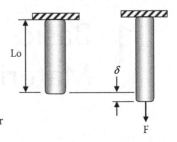

FIGURE 1.2 Bar object under axial tension (or compression) elongated.

Thus, the strain is defined as:

$$\varepsilon = \frac{\delta}{Lo} \text{ (unit in./in.)} \tag{1.2}$$

Starting with ductile material, where the material gradually ruptures, and before rupturing and after passing the elastic region, the material goes through the yielding phase, the stress-strain behavior of the material can be plotted as shown in Figure 1.3:

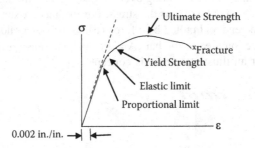

FIGURE 1.3 Stress-strain curve for typical aluminum material (ductile material).

The first point of interest on this curve is the proportional limit, where up to that point the behavior of the material is linear for stress vs. strain. The slope of this linear region is also known as the elastic modulus of the material (E). The next point of interest on this curve is the elastic limit point, where up to this point the material is perfectly elastic, and if the loading on the material is released, it will deform back into its original shape as if no loading was ever applied to the object. After this point, the material begins yielding, which at the yield point (where the corresponding stress for that location is known as the yield strength), the material has deformed permanently. The unloading of the material at this point does not constitute a return to the original shape.

It was indicated that the stress value corresponding to the stress where permanent strain is observed from the stress-strain curve data is known as the yield strength of the material (S_{yld}). This permanent strain value is usually taken at 0.2% (0.002 in./in.) of the original gage length of the specimen being tested for producing the stress-strain curve. (A straight line is drawn from there to the stress-strain curve, and the stress at that point is defined as the yield strength.) The next point of interest on the curve is the ultimate strength point where the highest capability of the material or the ultimate strength of the material is observed. Finally, fracture point is the point of the material rupture where the material gives in to the applied loading.

For brittle material where rupture is sudden and no gradual behavior of the stress vs. strain is observable before material fracture, all five (5) points of interest on the stress-strain curve occur at the same location (Figure 1.4).

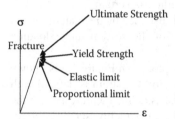

FIGURE 1.4 Stress-strain curve for typical brittle material (glass, cast iron, composites).

There are several fundamental terms that are known as the material "constants" derived from the testing of the materials normally in tensile tests. Starting with the definition given by Hooke's law, the stress and strain are related to each other by a material constant. Within the linear elastic region of the material this relationship is defined by the modulus of elasticity of the material (E). The modulus of elasticity is defined as the ratio of increment of unit stress (σ) to increment of the unit deformation (ε).

$$E = \frac{\sigma}{\varepsilon} \text{ (unit of psi)} \tag{1.3}$$

For isotropic homogeneous material, the strain deformation in axial direction is related to the lateral strain deformations in the other two directions. This relationship is defined by Poisson's ratio as:

$$v = -\frac{\varepsilon_y}{\varepsilon_x} = -\frac{\varepsilon_z}{\varepsilon_x} \tag{1.4}$$

Similar to modulus of elasticity, there is a modulus that relates the shear stress to the shearing strain of the material known as the shear modulus (G). Thus, this shear modulus can be defined as:

$$G = \frac{\tau}{\gamma} \text{ (unit of psi)} \tag{1.5}$$

Because there exists a relationship between the normal strain and the shear strain of the material, there must be a relationship between the modulus of elasticity and the shear modulus, which is normally derived as:

$$G = \frac{E}{2(1 + v)} \tag{1.6}$$

The ultimate strength of the material is defined as the maximum load sustained by the material per original cross-sectional area of the section under loading application (S_{ult}). At this loading, the material failure is certain.

Shear strength is defined as the maximum load sustained by the material in the shearing mode per original cross-sectional area of the section under loading application (S_{shr}).

The stress-strain behavior shown by the curve plotted in Figure 1.3 is known as the engineering stress vs. strain curve. In this plot, the strain is defined with respect to the original length (L = Lo). Also, stress is defined with respect to the original cross-section area (A = Ao). However, in actuality specifically for ductile material, the stress increases as the material goes through the yielding until it reaches the fracture or rupture. Material stress-strain behavior as a function of true stress vs. true strain, where true stress is the applied load over the actual instantaneous cross-sectional area (A≠Ao) and the true strain is the incremental change in length (ΔL) per corresponding length (L ≠ Lo), is shown in the following graph (Figure 1.5).

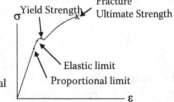

FIGURE 1.5 True stress-strain curve for typical aluminum material (ductile material).

Thus, the stress is defined as:

$$\sigma_t = \frac{F}{A} \text{ (unit of psi)} \tag{1.7}$$

Thus, the strain is defined as:

$$\varepsilon_t = \int_{Lo}^{L} \frac{1}{L} dL = In \cdot \frac{L}{Lo} \text{ (unit of in./in.)} \tag{1.8}$$

At locations where there are notches, holes, fillets and any other abrupt changes in the geometry, the stress values are higher than the normal stress distribution in the section under loading. These locations are known as the stress riser or stress concentration location. The ratio of the maximum stress at these locations to the nominal distributed stress at the far field is defined as the stress concentration factor (K_t), which is independent of the material.

The stress concentration factors for several geometries with axial and bending loadings are represented graphically by the following figures (Figures 1.6–1.10).

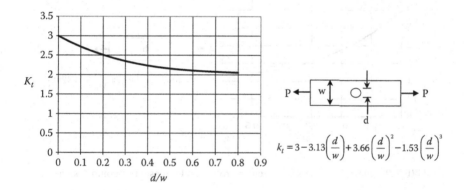

$$k_t = 3 - 3.13\left(\frac{d}{w}\right) + 3.66\left(\frac{d}{w}\right)^2 - 1.53\left(\frac{d}{w}\right)^3$$

FIGURE 1.6 Plate with transverse hole under axial tension or compression.

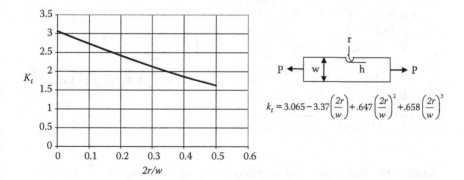

$$k_t = 3.065 - 3.37\left(\frac{2r}{w}\right) + .647\left(\frac{2r}{w}\right)^2 + .658\left(\frac{2r}{w}\right)^3$$

FIGURE 1.7 Plate with two semi-circular grooves under axial tension or compression.

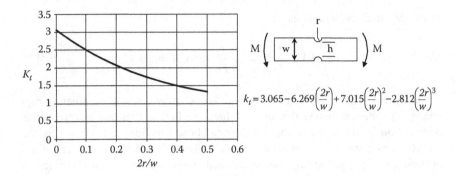

FIGURE 1.8   Plate with two semi-circular grooves under bending.

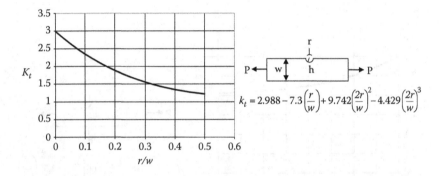

FIGURE 1.9   Plate with one semi-circular groove under tension or compression.

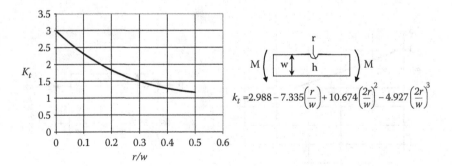

FIGURE 1.10   Plate with one semi-circular groove under bending.

It should be noted that, in brittle material the stress concentration causes the material failure if this stress is higher than the material ultimate strength. In ductile material, the stress concentration is higher than the yield strength of the material, which causes local plastic deformation.

The basics learned in this chapter are essential in carrying out stress analysis of metallic and composite materials. For analysis of the metallic materials since the materials are generally isotropic, only properties in one direction are needed. The materials possess the same properties in all and any direction the structural member is being analyzed. On the other hand, for composite materials since the materials are orthotropic, properties in major material coordinates are required to carry out the stress analysis of the structural members. In addition to that, since composites can be built to possess desired properties in any direction, the analysis may require more steps to develop the final stresses. Nevertheless, the lamina properties have to be known to develop the final lay-up properties.

Further, both the metallic and composite materials react to changes in the surrounding temperature. Thus, the material coefficient of thermal expansion, which defines these changes with respect to the temperature changes, has to be known for structural analysis. In addition to that, for stress analysis the material allowable can be knocked down for different temperature conditions. These knock-down factors can be determined from the available material data, usually presented in material handbooks. In the industry, usually every company has a set of material standards that are either derived from their own testing or adopted from the government testing of the commonly known materials. In the next section of this chapter, some of the properties for the commonly available materials used for structures are presented in tabular form. These values can be used for reference purposes to develop a general idea of the material properties. The metallic materials have several different strength levels depending on the type of alloys they have. Depending on the criticality of the actual design, the yield or ultimate strength of material can be selected as the design criteria.

1.3 THE MATERIAL PROPERTIES

Table 1.1 below illustrates typical material properties normally used for carrying out stress analysis of solids.

Table 1.2 below illustrates typical material properties normally used for carrying out stress analysis of composite laminates.

For understanding of the ductility of alloy materials, elongation limits of the materials can be examined for stress analysis purposes. Some of the typical elongation properties of alloys are illustrated in Table 1.3.

TABLE 1.1
Engineering Material Properties

Material	Modulus of Elasticity 10^6 psi	Modulus of Rigidity 10^6 psi	Poisson's Ratio lb/in^3	Specific Weight 10^{-6}	Coefficient of Thermal Expansion /°F	Ultimate Tensile Strength 10^3 psi	Ultimate Shear Strength 10^3 psi	Yield Strength 10^3 psi
STEEL								
ASTM-A36	29	11.2	0.3	0.284	6.5	58	–	36
ASTM-A242	29	11.2	0.3	0.284	6.5	70	–	50
ASTM-A441	29	11.2	0.3	0.284	6.5	67	–	46
ASTM-A572	29	11.2	0.3	0.284	6.5	60	–	42
ASTM-A514	29	11.2	0.3	0.284	6.5	110	–	100
Stainless Cold-rolled	28	10.8	0.3	0.286	9.6	125	–	75
Stainless Annealed	28	10.8	0.3	0.286	9.6	95	–	55
ALUMINUM								
2014-T4	10.6	4.1	0.3	0.101	12.8	62	38	42
2024-T4	10.6	4.1	0.3	0.101	12.9	68	41	47
6061-T6	10.0	3.7	0.3	0.098	13.1	45	30	40
7075-T6	10.4	3.9	0.3	0.101	12.0	83	48	73

COPPER								
Annealed	17	6.4	0.3	0.322	9.4	32	22	10
Hard-drawn	17	6.4	0.3	0.322	9.4	57	29	53
TITANIUM								
Alloy (6Al&4V)	16.5	–	0.3	0.161	5.3	130	–	120
PLASTIC								
Nylon	0.4	–	0.4	0.041	80	11	–	6.5
Polyester(TP)	0.35	–	0.4	0.048	75	8	–	–
Elastomer	0.03	–	0.4	0.043	–	6.5	–	–
Vinyl	0.45	–	0.4	0.052	75	6	–	–
RUBBER	10	4	0.5	0.033	4	2	–	–
GLASS	9.6	4.1	0.2	0.079	44	7	–	–
SILICON	23.9	–	0.22	0.084	–	–	–	1.02

TABLE 1.2
Composite Lamina Material Properties

Material	Modulus E_L Long 10^6 psi	Modulus E_T Transverse 10^6 psi	Modulus G_{LT} 10^6 psi	Poisson's Ratio v_{LT}	Density lb/in^3	Ultimate Tensile Strength 10^3 psi	Ultimate Tensile Strength 10^3 psi	Shear Strength 10^3 psi
IM6/Epoxy	29.4	1.62	1.22	0.32	0.058	508	8.1	14.2
Kevlar/Epoxy	12.6	0.8	0.32	0.34	0.050	186	4.4	7.1
T300/5208	26.3	1.49	1.04	0.28	0.058	218	5.8	9.9
T300/934	21.5	1.4	0.66	0.30	0.054	191	6.2	7.0
AS/3501	20.0	1.3	1.03	0.30	0.058	210	7.5	13.5
AS4/3501–6	20.6	1.49	1.04	0.27	0.057	331	8.3	10.3
S-glass/Epoxy	6.2	1.29	0.65	0.27	0.072	186	7.1	10.0
E-glass/Epoxy	5.7	1.25	0.55	0.28	0.076	157	5.7	12.9

TABLE 1.3
Alloy Material Elongation Properties

Alloy Material (%)	Percent Elongation
STEEL	
ASTM-A36	20
ASTM-A242	18
ASTM-A441	18
ASTM-A572	20
ASTM-A514	18
Stainless Cold-rolled	13
Stainless Annealed	13
ALUMINUM	
2014-T4	12
2024-T4	19
6061-T6	17
7075-T6	11
COPPER	
Annealed	50
Hard-drawn	50
TITANIUM	
Alloy (6Al&4V)	10

1.4   MATERIAL SELECTION

In the design failure analysis where loading is predefined, load path is preset, and the analysis shows that the stress levels are unacceptable; material substitution is a viable option for substantiation of the component under investigation. If the design has had previous material substitutions and/or various material selections, then the task of material selection is somewhat easier to carry out. In this case, the substation can be applied, and the failure analysis can be repeated to investigate the viability of the design. However, if the design is new or novel, this task requires a deeper look at the material selection process, and several factors and requirements need to be examined and established before a successful design is rendered. In this case, selection of the proper material for component design becomes an essential part of failure analysis. In material selection, the following aspects can be considered (Table 1.4).

For items 1, 2, 3 and 4, Tables 1.1 and 1.2 may be consulted to evaluate the material selection(s) for a specific design. For item 5, refer to Table 1.3 for some of the typical elongation properties of the alloy materials. For items 6 and 7, normally the Military Standard Handbook or MMPDS can be referred to as a good source. For items 8–12, usually the design applications have to be evaluated and the material vendors data need to be consulted to cover the design requirements.

TABLE 1.4

Material Selection Evaluation Aspects

Requirements	Evaluation Aspects
1. Strength/Volume Ratio	Ultimate and Yield Strength
2. Strength/Weight Ratio	Ultimate and Yield Strength
3. Stiffness	Elastic Modulus
4. Thermal Expansion	Coefficient of Thermal Expansion
5. Ductility	Elongation Limit
6. Toughness	Energy at Rapture
7. Temperature Resistance	Strength Reduction per Temperature
8. Corrosion Resistance	Operating Environment
9. Manufacturability	Process Specification/Strength
10. Cost	Cost per Weight
11. Wear and Tear	Operating Condition and Cycling
12. Availability	Procurement at the Market

Problems

1. Define in simple terms stress on a solid body and indicate the units of stress in SI and English systems.
2. Define strain and its relationship with stress in isotropic materials.
3. What is the material's Poisson's ratio and what is a typical value of Poisson's ratio for metallic material?
4. How can one determine the yield strength of a material from the stress and strain plot from the material tensile tests?
5. Take the average of the three metallic materials' ultimate tensile strengths and yield strengths and compare the ultimate tensile strengths to yield values.
6. In general, how do composite materials compare in strength to density ratio to the metallic material?
7. How do the elastic modulus compare to the shear modulus in typical steel materials and can one indicate a relationship?
8. For a flat plate with a circular hole in the middle, is the stress concentration higher in the close proximity to the hole or away from the hole?

REFERENCES

Beer, F.P., Johnston, E.R., DeWolf, J.T., *Mechanics of Material*, 2002. New York: MacGraw Hill.
Young, W.C., *Roark's Formulas for Stress & Strain*, 1989. New York: McGraw Hill.

2 Stress and Strain Relationship

2.1 INTRODUCTION

This chapter establishes the definitions of stress and strain. Further, Hooke's law relating stress to strain is discussed. The determination of principal stresses and strains is shown, and the application of Mohr's circle is elaborated. The stress equations applicable for use, in conjunction with Mohr's circle, are presented here, too. Hooke's law, defining the stress and strain relationships in a 3-D state of stress, are elaborated. Determination of plane stress and strain relationships are defined. The state of stress in polar coordinates is shown, and the concept of stress concentration around holes in plates is discussed. Theoretical concepts of strain gage measurements are defined and elaborated here. Numerical examples are provided to elaborate the concept and theory presented here.

2.2 STRESS

Stress is defined as application of a load over a finite area, where the area is an extension of a 3-D element. The state of stress is normally defined in terms of the stresses due to normal and shear loads acting on a structural body. It can be represented in either Cartesian or Polar coordinate systems; however, preference is given to a Cartesian system for general-shaped bodies. The state of stress on a body in three-dimensions is shown in Figure 2.1,

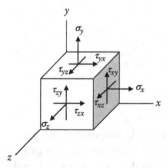

FIGURE 2.1 The 3-D state of stress.

where, σ_x, σ_y and σ_z are defined as normal stresses and τ_{xy}, τ_{yx}, τ_{yz}, τ_{zy}, τ_{xz} and τ_{zx} as the shear stresses present on a body. When the 3-D element is in equilibrium, the shear components reduce as following:

DOI: 10.1201/9781003311218-2

$$\tau_{xy} = \tau_{yx} \quad \tau_{yz} = \tau_{zy} \quad \tau_{zx} = \tau_{xz} \tag{2.1}$$

2.3   2-D STRESSES ON AN INCLINED ANGLE

The normal and shear stresses acting on a stress element at any inclined angle, shown by Figure 2.2, are calculated as following:

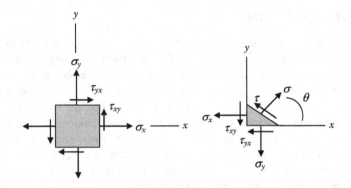

FIGURE 2.2   The 2-D state of stress for an inclined angle.

$$\sigma = \frac{\sigma_x + \sigma_y}{2} + \frac{\sigma_x - \sigma_y}{2} \cos 2\theta + \tau_{xy} \sin 2\theta \tag{2.2}$$

$$\tau = -\frac{\sigma_x - \sigma_y}{2} \sin 2\theta + \tau_{xy} \cos 2\theta \tag{2.3}$$

where the inclination angle θ is defined as:

$$\tan 2\theta = \frac{2\tau_{xy}}{\sigma_x - \sigma_y} \text{ or } 2\theta = \tan^{-1}\frac{2\tau_{xy}}{\sigma_x - \sigma_y} \tag{2.4}$$

2.4   PRINCIPAL STRESSES

Now, to determine the extreme poles of stress or (principal stresses), the equations (2.2) and (2.3) are differentiated with respect to θ and set to zero to determine the principal angle 2θ. Then, the principal angles are substituted back into equation (2.2) and (2.3) to determine the principal stresses.

$$\frac{d\sigma}{d\theta} = -(\sigma_x - \sigma_y)\sin 2\theta + 2\tau_{xy} \cos 2\theta = 0 \tag{2.5}$$

$$\tan 2\theta_p = \frac{2\tau_{xy}}{\sigma_x - \sigma_y} \text{ or } 2\theta_p = \tan^{-1}\frac{2\tau_{xy}}{\sigma_x - \sigma_y} \quad (2.6)$$

To substitute equation (2.6) back into equation (2.2), one would have the principal stresses as:

$$\sigma_{1,2} = \frac{\sigma_x + \sigma_y}{2} \pm \sqrt{\left(\frac{\sigma_x - \sigma_y}{2}\right)^2 + \tau_{xy}^2} \quad (2.7)$$

Likewise for maximum shear,

$$\frac{d\tau}{\partial\theta} = -2\left(\frac{\sigma_x - \sigma_y}{2}\right)\cos 2\theta - 2\tau_{xy}\sin 2\theta = 0 \quad (2.8)$$

$$\tan 2\theta_s = -\left(\frac{\sigma_x - \sigma_y}{2\tau_{xy}}\right) \text{ or } 2\theta_s = \tan^{-1} -\left(\frac{\sigma_x - \sigma_y}{2\tau_{xy}}\right) \quad (2.9)$$

To substitute equation (2.9) back into equation (2.3), one would have the maximum shear stresses as:

$$\tau_{1,2} = \pm\sqrt{\left(\frac{\sigma_x - \sigma_y}{2}\right)^2 + \tau_{xy}^2} \quad (2.10)$$

Now that the principal and maximum shear stress relations are developed, the graphical representation of them can be shown by Mohr's circle. Mohr's circle is used in determination of the state of stress of any point, at any plane direction. The construction of Mohr's circle is shown in Figure 2.3 and described as follows.

2.5 MOHR'S CIRCLE

Establish the coordinate system with x-axis representing the normal stresses and y-axis representing the shear stresses. Indicate the center of Mohr's circle by taking the average of the normal stress (C). Locate point A by using the stress element notations shown in Figure 2.2. The normal stress components in Figure 2.2 represent all positive states of stresses. Likewise, locate point B by its stress components. Draw a circle centered at point C with diameter AB. Draw the line from point A to point B.

Now, the stress state transformation can be determined graphically by rotating the line AB at the transformation plane angle. The principal stresses and the maximum shear stress can be determined by rotation of the line AB at the angles θ_p and θ_s, respectively.

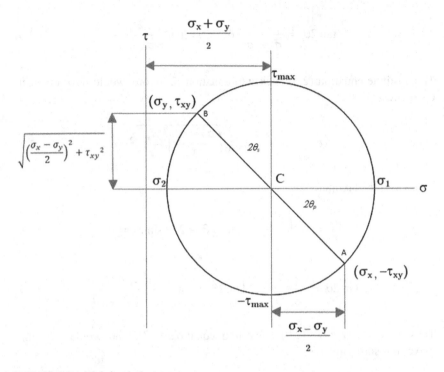

FIGURE 2.3 Mohr's circle.

Example 2.1: The element A is located on the top surface of the 1 in. shaft. The stress element A would be subjected to a moment (M) and Torque (T) due to a downward load of 1000 lb. Determine the principal stresses at that shaft location and draw Mohr's circle.

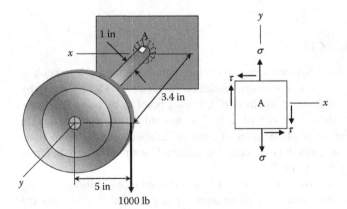

$$M = (3.4)(1000) = 3400 \ in - lb$$

$$\sigma = \frac{Mc}{I} \quad then, \quad \sigma = \frac{(3400)(0.5)}{0.049} = 35000 \ psi \ (I = \pi D^4/64)$$

$$T = (5)(-1000) = -5000 \ in - lb$$

$$\tau = \frac{Tc}{J} \quad then, \quad \tau = \frac{(-5000)(0.5)}{0.098} = -25510 \ psi \ (J = \pi D^4/32)$$

Knowing,

$$\sigma_{1,2} = \frac{\sigma_x + \sigma_y}{2} \pm \sqrt{\left(\frac{\sigma_x - \sigma_y}{2}\right)^2 + \tau_{xy}^2}, \text{ then the principal stresses are:}$$

$$\sigma_{1,2} = \frac{0 + 35000}{2} \pm \sqrt{\left(\frac{0 - 35000}{2}\right)^2 + (-25510)^2} = 48436 \ and$$

$$- 13436 \ psi$$

Using equation (2.6) the principal angle is:

$$2\theta_p = \tan^{-1}\frac{2(-25510)}{0 - 35000}, \quad thus \ \theta_p = 27.77°$$

Knowing,

$$\tau_{1,2} = \pm\sqrt{\left(\frac{\sigma_x - \sigma_y}{2}\right)^2 + \tau_{xy}^2}, \text{ then the maximum shear stresses are:}$$

$$\tau_{1,2} = \pm\sqrt{\left(\frac{0 - 35000}{2}\right)^2 + (-25500)^2} = \pm30936 \ psi$$

Using equation (2.9) the shear angle is,

$$2\theta_s = \tan^{-1} -\left(\frac{0 - 35000}{2(-25510)}\right) \quad thus, \ \theta_s = -17.23°$$

$$\sigma_{avg} = \frac{0 + 35000}{2} = 17500 \ psi$$

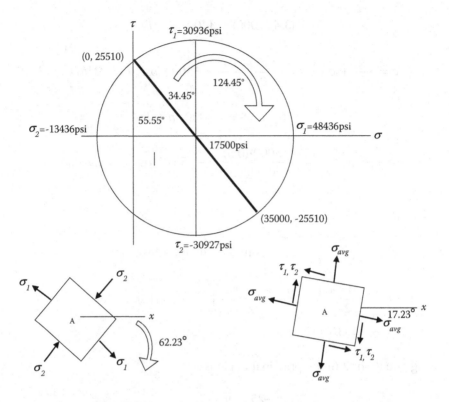

Example 2.1: (S.I. Units)

The element A is located on the top surface of the 1 in. shaft. The stress element A would be subjected to a moment (*M*) and Torque (*T*) due to a downward load of 4448.2 N. Determine the principal stresses at that shaft location and draw Mohr's circle.

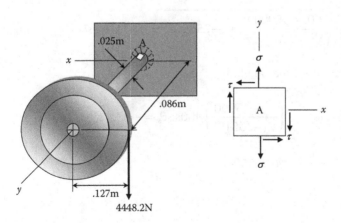

$$M = (.086)(4448.2) = 9.67E - 5N - m$$

$$\sigma = \frac{Mc}{I} \quad \text{then,} \quad \sigma = \frac{(9.67E - 5)(0.0125)}{1.92E - 8} = 241.3 \, MPa \quad (I = \pi D^4/64)$$

$$T = (.127)(-4448.2) = -564.9N - m$$

$$\tau = \frac{Tc}{J} \quad \text{then,} \quad \tau = \frac{(-564.9)(0.025)}{3.83E - 8} = -175.9 \, MPa \quad (J = \pi D^4/32)$$

Knowing,

$\sigma_{1,2} = \frac{\sigma_x + \sigma_y}{2} \pm \sqrt{\left(\frac{\sigma_x - \sigma_y}{2}\right)^2 + \tau_{xy}^2}$, then the principal stresses are:

$$\sigma_{1,2} = \frac{0 + 241.3}{2} \pm \sqrt{\left(\frac{0 - 241.3}{2}\right)^2 + (-175.9)^2} = 334 \, and - 92.7 \, MPa$$

Using equation (2.6) the principal angle is:

$$2\theta_p = \tan^{-1} \frac{2(-175.9)}{0 - 241.3}, \quad \text{thus } \theta_p = 27.77°$$

Knowing,

$\tau_{1,2} = \pm\sqrt{\left(\frac{\sigma_x - \sigma_y}{2}\right)^2 + \tau_{xy}^2}$, then the maximum shear stresses are:

$$\tau_{1,2} = \pm\sqrt{\left(\frac{0 - 241.3}{2}\right)^2 + 175.9^2} = \pm213.3 \, MPa$$

Using equation (2.9) the shear angle is,

$$2\theta_s = \tan^{-1} - \left(\frac{0 - 241.3}{2(-175.9)}\right) \quad \text{thus, } \theta_s = -17.23°$$

$$\sigma_{avg} = \frac{0 + 241.3}{2} = 120.7 \, MPa$$

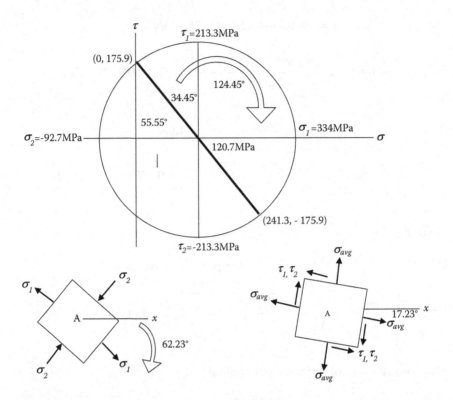

Example 2.2: For the state of stress shown below, determine the principal stresses and the maximum shear and draw Mohr's circle.

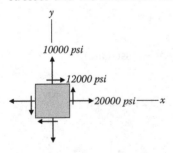

Knowing,

$$\sigma_{1,2} = \frac{\sigma_x + \sigma_y}{2} \pm \sqrt{\left(\frac{\sigma_x - \sigma_y}{2}\right)^2 + \tau_{xy}^2},$$ then the principal stresses are:

$$\sigma_{1,2} = \frac{20000 + 10000}{2} \pm \sqrt{\left(\frac{20000 - 10000}{2}\right)^2 + 12000^2}$$

$$= 28000 \ and \ 2000 \ psi$$

Using equation (2.6) the principal angle is:

$$2\theta_p = \tan^{-1}\frac{2(12000)}{20000 - 10000}, \quad \text{thus } \theta_p = 33.69°.$$

Knowing,

$\tau_{1,2} = \pm\sqrt{\left(\frac{\sigma_x - \sigma_y}{2}\right)^2 + \tau_{xy}^2}$, then the maximum shear stresses are:

$$\tau_{1,2} = \pm\sqrt{\left(\frac{20000 - 10000}{2}\right)^2 + 12000^2} = \pm13000 \ psi$$

Using equation (2.9) the shear angle is,

$$2\theta_s = \tan^{-1}\left(-\frac{20000 - 10000}{2(12000)}\right) \quad \text{thus,} \quad \theta_s = -11.31°.$$

$$\sigma_{avg} = \frac{20000 + 10000}{2} = 15000 \ psi$$

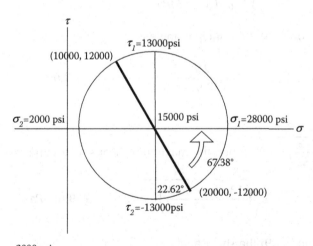

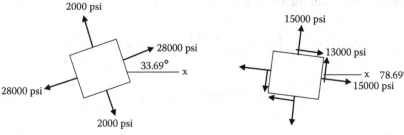

Example 2.2: (S.I. Units)

For the state of stress shown below, determine the principal stresses and the maximum shear and draw Mohr's circle.

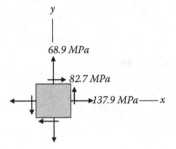

Knowing,

$$\sigma_{1,2} = \frac{\sigma_x + \sigma_y}{2} \pm \sqrt{\left(\frac{\sigma_x - \sigma_y}{2}\right)^2 + \tau_{xy}^2}, \text{ then the principal stresses are:}$$

$$\sigma_{1,2} = \frac{137.9 + 68.9}{2} \pm \sqrt{\left(\frac{137.9 - 68.9}{2}\right)^2 + 82.7^2}$$

$$= 193.1 \text{ and } 13.8 \text{ MPa}$$

Using equation (2.6) the principal angle is:

$$2\theta_p = \tan^{-1}\frac{2(82.7)}{137.9 - 68.9}, \quad \text{thus } \theta_p = 33.69°.$$

Knowing,

$$\tau_{1,2} = \pm\sqrt{\left(\frac{\sigma_x - \sigma_y}{2}\right)^2 + \tau_{xy}^2}, \text{ then the maximum shear stresses are:}$$

$$\tau_{1,2} = \pm\sqrt{\left(\frac{137.9 - 68.9}{2}\right)^2 + 82.7^2} = \pm89.6 \text{ MPa}$$

Using equation (2.9) the shear angle is,

$$2\theta_s = \tan^{-1} - \left(\frac{137.9 - 68.9}{2(82.7)}\right) \quad \text{thus, } \theta_s = -11.31°.$$

$$\sigma_{avg} = \frac{137.9 + 68.9}{2} = 103.4 \text{ MPa}$$

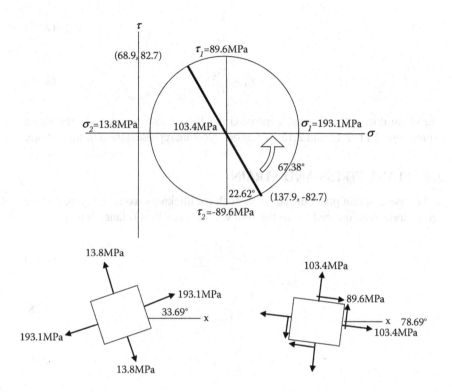

2.6 STRAIN

Strain is defined as the deformation in any direction per length. Based on the generalized Hooke's law, the corresponding strain in the body shown in Figure 2.1 (where the stresses do not exceed the proportional limit of the material) can be represented as:

$$\varepsilon_x = \frac{\sigma_x}{E} - v\frac{\sigma_y}{E} - v\frac{\sigma_z}{E} \tag{2.11}$$

$$\varepsilon_y = -v\frac{\sigma_x}{E} + \frac{\sigma_y}{E} - v\frac{\sigma_z}{E} \tag{2.12}$$

$$\varepsilon_z = -v\frac{\sigma_x}{E} - v\frac{\sigma_y}{E} + \frac{\sigma_z}{E} \tag{2.13}$$

$$\gamma_{xy} = \frac{\tau_{xy}}{G} \tag{2.14}$$

$$\gamma_{yz} = \frac{\tau_{yz}}{G} \qquad (2.15)$$

$$\gamma_{zx} = \frac{\tau_{zx}}{G} \qquad (2.16)$$

The strain due to the thermal expansion ($\alpha \Delta T$) can also be added to the strain expressions 2.11, 2.12 and 2.13 for a more accurate representation of total strain.

2.7 PLANE STRESS AND STRAIN

In the case of a thin plate, the stress through the thickness would be zero, and the stress-strain relations reduce to the following expressions (plane stress):

$$\sigma_x = E \frac{\varepsilon_x + \upsilon \varepsilon_y}{1 - \upsilon^2} \qquad (2.17)$$

$$\sigma_y = E \frac{\varepsilon_y + \upsilon \varepsilon_x}{1 - \upsilon^2} \qquad (2.18)$$

$$\varepsilon_z = -\frac{\upsilon}{1 - \upsilon}(\varepsilon_x + \varepsilon_y) \qquad (2.19)$$

To derive expressions (2.17) and (2.18) set $\sigma_z = 0$ in equations (2.11) and (2.12) and solve the two simultaneous equations for σ_x and σ_y. Then, to derive expression (2.19) substitute the two expressions (2.17) and (2.18) into equation (2.13).

For the case where the longitudinal movement of the body is constrained in the z-direction ($\varepsilon_z = 0$), the stress-strain relations reduce to the following expressions (plane stain):

$$\sigma_z = \upsilon(\sigma_x + \sigma_y) \qquad (2.20)$$

$$\varepsilon_x = \frac{1}{E}[(1 - \upsilon^2)\sigma_x - \upsilon(1 + \upsilon)\sigma_y] \qquad (2.21)$$

$$\varepsilon_y = \frac{1}{E}[(1 - \upsilon^2)\sigma_y - \upsilon(1 + \upsilon)\sigma_x] \qquad (2.22)$$

To derive the expression (2.20), start by setting the equation (2.13) to zero. To derive the expressions (2.21) and (2.22), substitute the equation (2.20) into equation (2.12) and (2.13), respectively.

2.8   PRINCIPAL STRAINS

In theory, the strain transformation of a point in a structural body is similar to stress transformations. Thus, the principal strains can be derived in the same manner. The principal strains of a point with respect to the two perpendicular directions and 'x' and 'y' are as follows:

$$\varepsilon_{1,2} = \frac{\varepsilon_x + \varepsilon_y}{2} \pm \frac{1}{2}\sqrt{(\varepsilon_x - \varepsilon_y)^2 + \gamma_{xy}^2} \tag{2.23}$$

Whereas the direction of the principal strain would be:

$$\tan 2\theta = \frac{\gamma_{xy}}{\varepsilon_x - \varepsilon_y} \tag{2.24}$$

Likewise, the maximum shear strain is shown as:

$$\left(\frac{\gamma}{2}\right)_{max} = \frac{1}{2}\sqrt{(\varepsilon_x - \varepsilon_y)^2 + \gamma_{xy}^2} \tag{2.25}$$

2.9   STRESS BASED ON THE MEASURED STRAINS

The stresses at a point on a structural surface can be determined by measurement of the strains at that point. This can be done by mounting strain gages on the surface of the structure and measuring the strains along any three line directions at some angle apart. Figure 2.4 illustrates this concept.

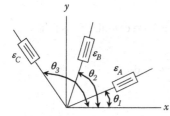

FIGURE 2.4   The strain gages rosette.

The normal strain in the direction of strain gages can be written in terms of the strain components ε_x, ε_y and γ_{xy} as follows:

$$\varepsilon_A = \varepsilon_x \cos^2 \theta_1 + \varepsilon_y \sin^2 \theta_1 + \gamma_{xy} \sin \theta_1 \cos \theta_1 \tag{2.26}$$

$$\varepsilon_B = \varepsilon_x \cos^2 \theta_2 + \varepsilon_y \sin^2 \theta_2 + \gamma_{xy} \sin \theta_2 \cos \theta_2 \tag{2.27}$$

$$\varepsilon_C = \varepsilon_x \cos^2 \theta_3 + \varepsilon_y \sin^2 \theta_3 + \gamma_{xy} \sin \theta_3 \cos \theta_3 \qquad (2.28)$$

Example 2.3:

The strains at a point are measured using a 45° rosette. The strains measured are $\varepsilon_A = 40\ \mu$, $\varepsilon_B = 1000\ \mu$ and $\varepsilon_C = 400\ \mu$. Determine the 'x' and 'y' strain components for normal and shear strains. Calculate the normal stresses in 'x' and 'y' direction and the shear stress. Also calculate the principal strains.

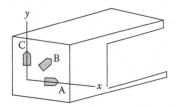

By relations 2.26–2.28,

$$40 = \varepsilon_x \cos^2 0 + \varepsilon_y \sin^2 0 + \gamma_{xy} \sin 0 \cos 0 \qquad\qquad 40 = \varepsilon_x$$

$$1000 = \varepsilon_x \cos^2 45 + \varepsilon_y \sin^2 45 + \gamma_{xy} \sin 45 \cos 45 \quad 1000 = \varepsilon_x(0.5) + \varepsilon_y(0.5) + \gamma_{xy}(0.5)$$

$$400 = \varepsilon_x \cos^2 90 + \varepsilon_y \sin^2 90 + \gamma_{xy} \sin 90 \cos 90 \qquad 400 = \varepsilon_y$$

Solving the above relations simultaneously, one would have:

$$\varepsilon_x = 40\mu, \quad \varepsilon_y = 400\mu, \quad \text{and } \gamma_{xy} = 1560\mu$$

Using equations (2.23) and (2.25), the principal strains are calculated as follows:

$$\varepsilon_{1,2} = \frac{\varepsilon_x + \varepsilon_y}{2} \pm \frac{1}{2}\sqrt{(\varepsilon_x - \varepsilon_y)^2 + \gamma_{xy}^2} \text{ thus,}$$

$$\varepsilon_{1,2} = \frac{40 + 400}{2} \pm \frac{1}{2}\sqrt{(40 - 400)^2 + 1560^2} = 1020\mu \text{ and} - 580.5\mu$$

$$\left(\frac{\gamma}{2}\right)_{\max} = \frac{1}{2}\sqrt{(\varepsilon_x - \varepsilon_y)^2 + \gamma_{xy}^2} \text{ thus,}$$

$$\left(\frac{\gamma}{2}\right)_{\max} = \frac{1}{2}\sqrt{(40 - 400)^2 + 1560^2} = 800.5\mu$$

To determine the stresses, use the equations of plane strain, equations (2.21) and (2.22). For shear stress, use equation (2.14).

$$\varepsilon_x = \frac{1}{E}\left[(1 - v^2)\sigma_x - v(1 + v)\sigma_y \right]$$

$$\varepsilon_y = \frac{1}{E}\left[(1 - v^2)\sigma_y - v(1 + v)\sigma_x \right]$$

$$\gamma_{xy} = \frac{\tau_{xy}}{G}$$

2.10 STRESS STATE IN POLAR COORDINATES

To evaluate the state of stress in a polar coordinate system, consider a stress element in polar coordinates as shown in Figure 2.5.

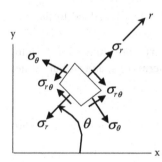

FIGURE 2.5 The stress element in polar coordinates.

The stresses in polar coordinates in terms of the stresses in the Cartesian coordinate system can be written as:

$$\sigma_r = \sigma_x \cos^2\theta + \sigma_y \sin^2\theta + 2\tau_{xy} \sin\theta \cos\theta \qquad (2.29)$$

$$\sigma_\theta = \sigma_x \sin^2\theta + \sigma_y \cos^2\theta - 2\tau_{xy} \sin\theta \cos\theta \qquad (2.30)$$

$$\tau_{r\theta} = (\sigma_y - \sigma_x)\sin\theta \cos\theta + \tau_{xy}(\cos^2\theta - \sin^2\theta) \qquad (2.31)$$

According to the generalized Hooke's law, the stress-strain relationship can be represented as the following in the polar coordinate system:

$$\sigma_r = \frac{E}{(1+v)(1-2v)}((1-v)\varepsilon_r + v\varepsilon_\theta + v\varepsilon_z) \tag{2.32}$$

$$\sigma_\theta = \frac{E}{(1+v)(1-2v)}(v\varepsilon_r + (1-v)\varepsilon_\theta + v\varepsilon_z) \tag{2.33}$$

$$\sigma_z = \frac{E}{(1+v)(1-2v)}(v\varepsilon_r + v\varepsilon_\theta + (1-v)\varepsilon_z) \tag{2.34}$$

$$\gamma_{r\theta} = \frac{\tau_{r\theta}}{G} \tag{2.35}$$

$$\gamma_{rz} = \frac{\tau_{rz}}{G} \tag{2.36}$$

$$\gamma_{z\theta} = \frac{\tau_{z\theta}}{G} \tag{2.37}$$

To add the thermal expansion effects, the term, $\frac{E\alpha\Delta T}{1-2v}$ can be added to the expressions (2.32), (2.33) and (2.34).

To develop the plane stress condition expressions, substitute $\sigma_z = 0$, into equation (2.34) and solve for the expression for ε_z in terms of ε_r and ε_θ. Substitute this term into equations (2.32) and (2.33).

$$\sigma_r = E\frac{\varepsilon_r + v\varepsilon_\theta}{1-v^2} \tag{2.38}$$

$$\sigma_\theta = E\frac{\varepsilon_\theta + v\varepsilon_r}{1-v^2} \tag{2.39}$$

Likewise, to develop the plane strain condition expression, substitute $\varepsilon_z = 0$ into equations (2.32) and (2.33). Then, solve the two equations (2.32) and (2.33) simultaneously for ε_r and ε_θ. The following expressions are developed for plane strain in a polar coordinate system.

$$\varepsilon_r = \frac{1}{E}[(1-v^2)\sigma_r - v(1+v)\sigma_\theta] \tag{2.40}$$

$$\varepsilon_\theta = \frac{1}{E}[(1-v^2)\sigma_\theta - v(1+v)\sigma_r] \tag{2.41}$$

2.11   STRESS FIELD AROUND CIRCULAR HOLES IN THIN PLATES

Considering the thin plate shown in Figure 2.6 with a small circular hole of radius $r = a$, one can derive the stress field distribution using the Airy stress function as follows.

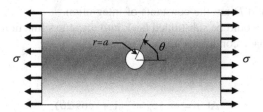

FIGURE 2.6   The large thin plate with small circular hole.

Starting with the compatibility equation for stress function $\phi = \phi(r, \theta)$ in polar coordinates, one would have:

$$\nabla^4 \phi = \left(\frac{\partial^2}{\partial r^2} + \frac{1}{r} \frac{\partial}{\partial r} + \frac{1}{r^2} \frac{\partial^2}{\partial \theta^2} \right) \left(\frac{\partial^2 \phi}{\partial r^2} + \frac{1}{r} \frac{\partial \phi}{\partial r} + \frac{1}{r^2} \frac{\partial^2 \phi}{\partial \theta^2} \right) = 0 \quad (2.42)$$

The stress components in the this polar coordinate system are given by

$$\sigma_r = \frac{1}{r} \frac{\partial \phi}{\partial r} + \frac{1}{r^2} \frac{\partial^2 \phi}{\partial \theta^2} \quad (2.43)$$

$$\sigma_\theta = \frac{\partial^2 \phi}{\partial r^2} \quad (2.44)$$

$$\tau_{r\theta} = \frac{1}{r^2} \frac{\partial \phi}{\partial \theta} - \frac{1}{r} \frac{\partial^2 \phi}{\partial r \partial \theta} \quad (2.45)$$

Assuming a stress function,

$$\phi = f(r)\cos 2\theta \quad (2.46)$$

and substituting this stress function into equation (2.42), one would develop:

$$\left(\frac{d^2}{dr^2} + \frac{1}{r} \frac{d}{dr} - \frac{4}{r^2} \right) \left(\frac{\partial^2 f}{\partial r^2} + \frac{1}{r} \frac{\partial f}{\partial r} - \frac{4f}{r^2} \right) = 0 \quad (2.47)$$

This differential equation would have a solution in the form:

$$f(r) = Ar^2 + Br^4 + C\frac{1}{r^2} + D \qquad (2.48)$$

where A, B, C and D are the constants of integration.

This solution would be substituted back into the stress function (2.46) and the following boundary conditions applied.

$$\text{At } r = a \quad \sigma_r = \tau_{r\theta} = 0 \qquad (2.49)$$

$$\text{At } r = \infty, \quad \sigma_r = \frac{\sigma}{2}(1 + \cos 2\theta), \qquad (2.50)$$

$$\sigma_\theta = \frac{\sigma}{2}(1 - \cos 2\theta) \qquad (2.51)$$

$$\tau_{r\theta} = \frac{\sigma}{2} \sin 2\theta \qquad (2.52)$$

Based on this operation, the constants of the integration for equation (2.48) are determined, and they are as follows:

$$A = -\frac{\sigma}{4}, \quad B = 0, \quad C = -\frac{a^2\sigma}{4}, \quad D = \frac{a^2\sigma}{2} \qquad (2.53)$$

Thus, the expression (2.46) would have the following final form:

$$\phi = \left(-\frac{\sigma r^2}{4} - \frac{a^4\sigma}{4r^2} + \frac{a^2\sigma}{2}\right)\cos 2\theta \qquad (2.54)$$

By substitution of this stress function into expressions (2.43), (2.44) and (2.45), one would have the following expression for stress distribution on a large plate containing a circular hole.

$$\sigma_r = \frac{\sigma}{2}\left[\left(1 - \frac{a^2}{r^2}\right) + \left(1 + \frac{3a^4}{r^4} - \frac{4a^2}{r^2}\right)\cos 2\theta\right] \qquad (2.55)$$

$$\sigma_\theta = \frac{\sigma}{2}\left[\left(1 + \frac{a^2}{r^2}\right) - \left(1 + \frac{3a^4}{r^4}\right)\cos 2\theta\right] \qquad (2.56)$$

$$\tau_{r\theta} = -\frac{\sigma}{2}\left(1 - \frac{3a^4}{r^4} + \frac{2a^2}{r^2}\right)\sin 2\theta \qquad (2.57)$$

The extreme stress values at $r = a$ at the edge of the circular hole are characterized as:

$$\sigma_{\theta\,max} = 3\sigma \quad at \quad \theta = \pm\frac{\pi}{2} \tag{2.58}$$

$$\sigma_{\theta\,min} = -\sigma \quad at \quad \theta = 0, \quad \theta = \pm\pi \tag{2.59}$$

Indicating a maximum stress of three times the nominal stress, which relates to a stress concentration factor of k = 3.

2.12 STRESS DUE TO THERMAL EXPANSION AND AXIAL PRE-LOADING

Consider a member under a tensile or compressive preload condition in combination with thermal load acting on it. Usually during installation of a member, a compressive or tensile load can be acting on the part when either bolting or fitting the part in place. Imagine a flanged-shaft being installed between two brackets in an assembly, as shown by Figure 2.7.

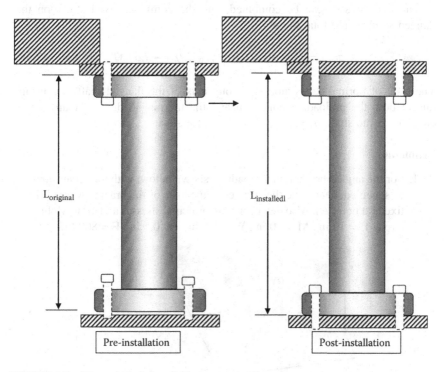

FIGURE 2.7   Flanged-shaft installation.

After such installation, there exist a preload tensile load on the shaft as the shaft is secured between the brackets by tightening the bolts. The axial stress on the shaft can be calculated by use of the delta elongation (ΔL) value of the shaft after installation as follows:

$$\sigma_{axial} = E\ \varepsilon_{axial} \tag{2.60}$$

or

$$\sigma_{axial} = E\,(\Delta L/L) \tag{2.61}$$

where $\Delta L = L_{installed} - L_{original}$ and $L = L_{original}$.

After installation, if the shaft member is under a temperature drop or temperature rise, then the thermal stress on the shaft member would be:

$$\sigma_{thermal} = E\ \varepsilon_{thermal} \tag{2.62}$$

or

$$\sigma_{thermal} = E\alpha\,(DT) \tag{2.63}$$

where ΔT = Change in temperature and α = Coefficient of thermal expansion.

The two stresses can be combined, and the combined axial stress on the flanged shaft would total to be:

$$\sigma_{total} = \sigma_{axial} - \sigma_{thermal} = E\,(\Delta L/L) - E\alpha\,(\Delta T) \tag{2.64}$$

The state of normal stress along the long axis of the flanged-shaft is constant due to both the preloading condition and the thermal loading. For any stress element on the shaft, $\sigma_x = \sigma_{total}$, $\sigma_y = 0$ and for shear $\tau_{xy} = 0$.

Problems

1. For the angled bracket under loading, shown below, with the dimensions as specified, determine the stresses at the base of the bracket where it is fixed (at point A). Also determine the principal stresses and draw Mohr's circle. L = 10 in., M = 10 in., W = 1.5 in., t = 0.5 in., F = 800 lbf.

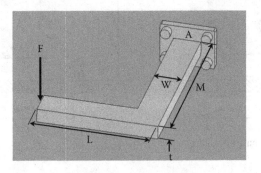

2. For the stress element shown below, determine the principal stresses and the corresponding Mohr's circle.

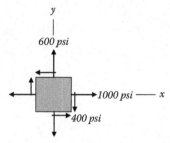

3. For the stress element shown below, determine the principal stresses and the corresponding Mohr's circle.

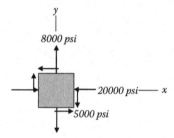

4. For the stress element shown below, determine the principal stresses and the corresponding Mohr's circle.

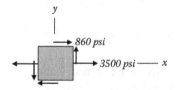

5. For the stress element shown below, determine the principal stresses and the corresponding Mohr's circle.

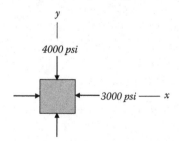

6. For the stress element shown below, determine the principal stresses and the corresponding Mohr's circle.

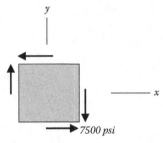

7. For the structure shown below, determine the principal stresses at locations A, B, C.

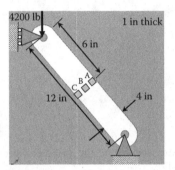

8. The strain measurement data from the 45° rosette shown below are given as,

$$\varepsilon_1 = 800 \ \mu\varepsilon, \quad \varepsilon_2 = 210 \ \mu\varepsilon, \quad \varepsilon_1 = 640 \ \mu\varepsilon,$$

Determine the corresponding strains in the x and y direction. Also, determine the principal strains.

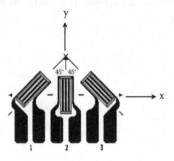

9. For the 0.05 in. thick rectangular plate with a 0.5 in. diameter hole at the center, plot the radial and angular stresses starting from the hole edge, at point A, to the plate edge, at point B.

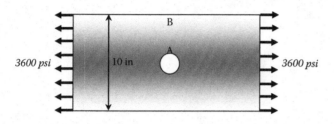

REFERENCES

Beer, F.P., Johnston, E.R., DeWolf, J.T., *Mechanics of Material*, 2002. New York: MacGraw Hill.
Timoshenko, S.P., Goodier, J.N., *Theory of Elasticity*, 1970. New York: McGraw Hill.
Ugural, A.C., Fenster, S.K., *Advanced Strength and Applied Elasticity*, 1995. New Jersey: Prentice Hall.

3 Stresses due to Various Loading Conditions

3.1 INTRODUCTION

In this chapter topics of contact stresses, pressure vessels and stresses due to various combined loadings are discussed. The contact stresses between two elastic bodies are seen in many systems such as ball bearings, trunnions and wheels on a railroad track. The maximum contact stress is directly normal to the plane of the contact area zone and is shown in this chapter for spheres and cylinders in contact. The next topic discussed is the stresses on the walls of the pressure vessels. Both thin-walled and thick-walled vessels are discussed. Also, the state of stress under various loading conditions is examined.

3.2 CONTACT STRESSES

The application of load over a small finite area between two elastic bodies pressing against each other is known as the concept of contact stress. Consider the sphere contacting an elastic solid platform shown in Figure 3.1. The contact zone is in circular form with a radius of "a". The sphere has a diameter of "d_1". The applied load on the sphere pressing down the sphere to the solid platform is P.

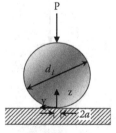

FIGURE 3.1 Sphere on elastic solid contact.

The maximum contact pressure between the sphere and the platform is known as the following:

$$P_{\max} = (0.919) \sqrt[3]{\frac{P}{d_1{}^2 \left(\frac{1 - v_1^2}{E_1} + \frac{1 - v_2^2}{E_2} \right)^2}} \tag{3.1}$$

DOI: 10.1201/9781003311218-3

Also, the radius of the contact zone is defined as

$$a = (0.721) \sqrt[3]{Pd_1\left(\frac{1 - v_1^2}{E_1} + \frac{1 - v_2^2}{E_2}\right)} \quad (3.2)$$

where E_1 is the elastic modulus of the sphere, E_2 is the elastic modulus of the solid platform, v_1 is the sphere's Poisson's ratio and v_2 is the solid platform's Poisson's ratio.

The contact stresses in x, y and z directions due to the maximum contact pressure are determined as follows for spherical contacts:

$$\sigma_x = \sigma_y = -Pmax\left[\left[1 - \left|\frac{z}{a}\right|tan^{-1}\left(\frac{1}{\left|\frac{z}{a}\right|}\right)\right](1 + v) - \frac{1}{2\left(1 + \left(\frac{z}{a}\right)^2\right)}\right] \quad (3.3)$$

$$\sigma_z = \frac{-Pmax}{1 + \left(\frac{z}{a}\right)^2} \quad (3.4)$$

Now consider a sphere contacting another sphere, as shown in Figure 3.2. The contact zone is in circular form again with a radius of "a". The spheres have diameters of "d_1" and "d_2". The applied load on the spheres pressing the spheres together is P.

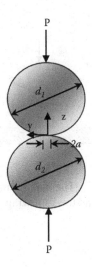

FIGURE 3.2 Sphere on sphere contact.

The maximum contact pressure between spheres is known as the following:

$$P_{max} = (0.919) \sqrt[3]{\frac{P}{\left(\frac{d_1 d_2}{d_1 + d_2}\right)^2 \left(\frac{1 - v_1^2}{E_1} + \frac{1 - v_2^2}{E_2}\right)^2}} \tag{3.5}$$

Also, the radius of contact zone is defined as

$$a = (0.721) \sqrt[3]{P\left(\frac{d_1 d_2}{d_1 + d_2}\right)\left(\frac{1 - v_1^2}{E_1} + \frac{1 - v_2^2}{E_2}\right)} \tag{3.6}$$

where E_1 is the elastic modulus of sphere 1, E_2 is the elastic modulus of sphere 2, v_1 is the Poisson's ratio for sphere 1 and v_2 is the Poisson's ratio for sphere 2.

The contact stresses in x, y and z directions due to the maximum contact pressure are determined as follows for spherical contacts:

$$\sigma_x = \sigma_y = -Pmax\left[\left[1 - \left|\frac{z}{a}\right| tan^{-1}\left(\frac{1}{\left|\frac{z}{a}\right|}\right)\right](1 + v) - \frac{1}{2\left(1 + \left(\frac{z}{a}\right)^2\right)}\right] \tag{3.7}$$

$$\sigma_z = \frac{-Pmax}{1 + \left(\frac{z}{a}\right)^2} \tag{3.8}$$

Also, consider a cylinder contacting an elastic solid platform, as shown in Figure 3.3. The contact zone is in rectangular form with a width of "b". The length of the contact zone is "L". The cylinder has a diameter of "d_1". The applied load on the cylinder pressing down the cylinder to the solid platform is P.

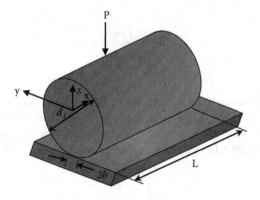

FIGURE 3.3 Cylinder on elastic solid contact.

The maximum contact pressure between the cylinder and the platform is known as the following:

$$P_{max} = (0.798) \sqrt{\frac{P}{Ld_1 \left(\frac{1 - v_1^2}{E_1} + \frac{1 - v_2^2}{E_2} \right)}} \tag{3.9}$$

Also, the width of the contact zone is defined as

$$b = (0.798) \sqrt{\frac{Pd_1}{L} \left(\frac{1 - v_1^2}{E_1} + \frac{1 - v_2^2}{E_2} \right)} \tag{3.10}$$

where E_1 is the elastic modulus of the cylinder, E_2 is the elastic modulus of the solid platform, v_1 is the Poisson's ratio for cylinder and v_2 is the Poisson's ratio of the solid platform.

The contact stresses in x, y and z directions due to the maximum contact pressure are determined as the following for cylindrical contacts:

$$\sigma_x = -2vPmax \left[\sqrt{1 + \left(\frac{z}{b} \right)^2} - \left| \frac{z}{b} \right| \right] \tag{3.11}$$

$$\sigma_y = -Pmax \left[\left(\frac{1 + 2\left(\frac{z}{b} \right)^2}{\sqrt{1 + \left(\frac{z}{b} \right)^2}} \right) - 2 \left| \frac{z}{b} \right| \right] \tag{3.12}$$

$$\sigma_z = \frac{-Pmax}{\sqrt{1 + \left(\frac{z}{b} \right)^2}} \tag{3.13}$$

Finally, consider two cylinders contacting each other, as shown in Figure 3.4. The contact zone is in rectangular form with a width of "b" again. The length of the contact zone is "L". The cylinders have diameters of "d_1" and "d_2". The applied load on the cylinders pressing the cylinders together is P.

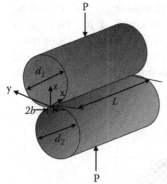

FIGURE 3.4 Cylinder on cylinder contact.

The maximum contact pressure between the cylinders is known as the following:

$$P_{\max} = (0.798)\sqrt{\frac{P}{L\left(\frac{d_1 d_2}{d_1 + d_2}\right)\left(\frac{1-v_1^2}{E_1} + \frac{1-v_2^2}{E_2}\right)}} \qquad (3.14)$$

Also, the width of contact zone is defined as

$$b = (0.798)\sqrt{\frac{P}{L}\left(\frac{d_1 d_2}{d_1 + d_2}\right)\left(\frac{1-v_1^2}{E_1} + \frac{1-v_2^2}{E_2}\right)} \qquad (3.15)$$

where E_1 is the elastic modulus of cylinder 1, E_2 is the elastic modulus of cylinder 2, v_1 is the Poisson's ratio for cylinder1 and v_2 is the Poisson's ratio of the cylinder 2.

The contact stresses in x, y and z directions due to the maximum contact pressure are determined as follows for cylindrical contacts:

$$\sigma_x = -2vP max\left[\sqrt{1 + \left(\frac{z}{b}\right)^2} - \left|\frac{z}{b}\right|\right] \qquad (3.16)$$

$$\sigma_y = -P max\left[\left(\frac{1 + 2\left(\frac{z}{b}\right)^2}{\sqrt{1 + \left(\frac{z}{b}\right)^2}}\right) - 2\left|\frac{z}{b}\right|\right] \qquad (3.17)$$

$$\sigma_z = \frac{-P max}{\sqrt{1 + \left(\frac{z}{b}\right)^2}} \qquad (3.18)$$

Example 3.1: A railway car with wheels of diameter 31 in. is rolling over a railroad track. The width of the car wheels is 4 in. The car is applying a compression load of 44,000 lbf on the railway. Determine the maximum contact pressure between the railway and the car wheel. Assume steel wheels and steel railroad tracks ($E = 29 \times 10^3$ psi and $v = 0.3$).

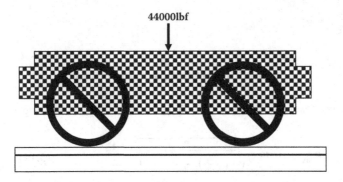

The maximum contact pressure between the wheel and the railroad track is determined by equation (3.5),

$$P_{max} = (0.798) \sqrt[3]{\frac{P}{Ld_1\left(\frac{1 - v_1^2}{E_1} + \frac{1 - v_2^2}{E_2}\right)}}$$

where $d_1 = 31$ in., $L = 4$ in. and $P = 22,000$ lbf. Note: $P = 44,000/2 = 22,000$ lbf per wheel

$$P_{max} = (0.798) \sqrt[3]{\frac{22000}{(4)(31)\left(\frac{1 - 0.3^2}{29 \times 10^6} + \frac{1 - 0.3^2}{29 \times 10^6}\right)}} = 1125.5 psi$$

3.3   THIN-WALLED SPHERICAL PRESSURE VESSELS

Consider a spherical pressure vessel under internal gage pressure p. Due to its symmetrical geometry, the normal stresses on the vessel walls in all directions would be equal. There would be no shear stress on the walls. To illustrate the state of stress on the surface of this type of vessel, the vessel can be cut at any cross-section, and the free-body diagram can be derived. Figure 3.5 below illustrates this free-body diagram.

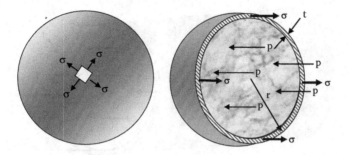

FIGURE 3.5 Cross-section of the spherical pressure vessel.

The force balance is

$$\sum F = 0 \quad \sigma 2\pi rt = p\pi r^2 \tag{3.19}$$

Thus, by rearranging the terms,

$$\sigma = \frac{p\pi r^2}{2\pi rt} \text{ or } \sigma = \frac{pr}{2t} \tag{3.20}$$

3.4 THIN-WALLED CYLINDRICAL PRESSURE VESSELS

Now consider a cylindrical pressure vessel under an internal gage pressure. There would be two stress components in the plane of the vessel cross-section, the longitudinal stress σ_{long} and the hoop stress σ_{hoop} (Figure 3.6).

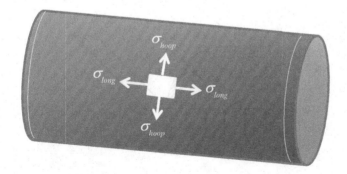

FIGURE 3.6 Cylindrical pressure vessel.

When the vessel is under static equilibrium, the free-body diagram shown below in Figure 3.7 can be used to derive the stress component in the longitudinal direction.

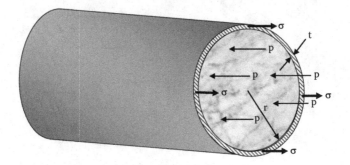

FIGURE 3.7 Cylindrical vessel cross-section.

The force balance is

$$\sum F = 0 \quad \sigma_{long} 2\pi r t = p\pi r^2 \tag{3.21}$$

Thus, by rearranging the terms,

$$\sigma_{long} = \frac{p\pi r^2}{2\pi r t} \quad \text{or} \quad \sigma_{long} = \frac{pr}{2t} \tag{3.22}$$

This next free-body diagram is used to derive the hoop stress on the vessel wall (Figure 3.8).

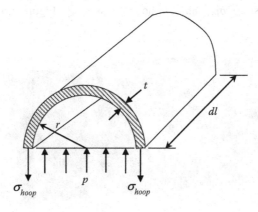

FIGURE 3.8   The cylindrical vessel cross-section.

$$\sum F = 0 \quad \sigma_{hoop} 2t\,dl = p2r\,dl \tag{3.23}$$

Thus,

$$\sigma_{hoop} = \frac{p2rdl}{2tdl} \text{ or } \sigma_{hoop} = \frac{pr}{t} \qquad (3.24)$$

Example 3.2: For a steel air pressure vessel that is 84 in. long and 35 in. in diameter with wall thickness of 0.375 in., determine the longitudinal and hoop stresses on the pressure vessel walls. Compare the stress results with the steel allowable to determine a safety factor. The internal air pressure inside the vessel is 200 psi.

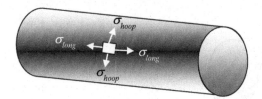

Using equations (3.22) and (3.24), one would have

$$\sigma_{long} = \frac{pr}{2t} \text{ thus,}$$

$$\sigma_{long} = \frac{200(35)}{2(.375)} = 9333psi$$

$$\sigma_{hoop} = \frac{pr}{t} \text{ thus,}$$

$$\sigma_{hoop} = \frac{200(35)}{.375} = 18670psi$$

The steel has a yield strength of 36000 psi; thus, the safety factor is calculated as

$$\sigma_{yld} = 36000\,psi$$

$$\sigma_{max} = \sigma_{hoop} = 18670\,psi$$

Then,

$$F.S. = \frac{\sigma_{yld}}{\sigma_{max(hoop)}} = \frac{36000}{18670} = 1.93$$

Example 3.2: (S.I. Units)

For a steel air pressure vessel that is 2.13 m long and 0.89 m in diameter with wall thickness of 9.53E-3 m, determine the longitudinal and hoop stresses on the pressure vessel walls. Compare the stress results with the steel allowable to determine a safety factor. The internal air pressure inside the vessel is 1.380 MPa.

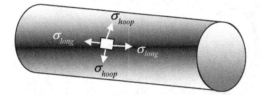

Using equations (3.22) and 3.24(3.14), one would have

$$\sigma_{long} = \frac{pr}{2t} \text{ thus,}$$

$$\sigma_{long} = \frac{1.38E6(.89)}{2(9.53E-3)} = 64.35 \, MPa$$

$$\sigma_{hoop} = \frac{pr}{t} \text{ thus,}$$

$$\sigma_{hoop} = \frac{1.38E6(.89)}{9.53E-3} = 128.7 \, MPa$$

The steel has a yield strength of 248.2 MPa; thus, the safety factor is calculated as

$$\sigma_{yld} = 248.2 \, MPa$$

$$\sigma_{yld} = 248.2 \, MPa$$

$$\sigma_{max} = \sigma_{hoop} = 128.7 \, MPa$$

Then,

$$F.S. = \frac{\sigma_{yld}}{\sigma_{max(hoop)}} = \frac{248.2}{128.7} = 1.93$$

Example 3.3: A 2-in. wide steel cylindrical ring is fitted over a 2-in. wide aluminum cylindrical ring, as shown by the following figure. At room temperature, the steel ring has an exact inner diameter of 4 in. and is 0.125 in. thick. Also at room temperature, the aluminum ring has an exact outer diameter of 4 in. and is 0.25 in. thick. The temperature of the fitting is raised to 130°F. Determine the stress on the steel ring and the pressure exerted on the steel ring by the aluminum ring.

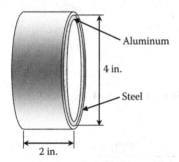

Aluminum

4 in.

Steel

2 in.

It should be noted that because the two rings are originally fitted exactly over each other and they are at equilibrium at the same temperature, then the steel cylinder's total circumferential elongation is equal to the aluminum cylinder's circumferential elongation after the temperature rise.

Now, the change in the circumferential length of the rings over the original ring circumference is the total strain the ring sees. Thus, the circumferential elongation, $\Delta elong$, can be defined as

$$\Delta elong = 2\pi r \varepsilon_{total}$$

The total strain for the steel cylinder is calculated as the strain due to thermal expansion plus the strain due to the pressure exerted by the aluminum ring expansion,

$$\varepsilon_{total,steel} = \varepsilon_{steel,T} + \varepsilon_{steel,P}$$

where $\varepsilon_{steel,T} = \alpha_{steel}\Delta T$ and $\varepsilon_{steel,P} = \dfrac{\sigma_{steel}}{E_{steel}} = \dfrac{\frac{Pr}{t_{steel}}}{E_{steel}}$

Knowing the steel properties, $E_{steel} = 29 \times 10^6 psi$ and $\alpha_{steel} = 6.5 \times 10^{-6}/°F$, then

$$\varepsilon_{total,steel} = 6.5 \times 10^{-6}(130 - 75) + \frac{P(2)}{29 \times 10^6(0.125)}$$

and the steel ring's elongation is

$$\Delta elong_{steel} = 2\pi\frac{4}{2}\left(6.5 \times 10^{-6}(130 - 75) + \frac{P(2)}{29 \times 10^6(0.125)}\right).$$

Similarly, for the aluminum ring the total strain is

$$\varepsilon_{total,AL} = \varepsilon_{AL,T} + \varepsilon_{AL,P}$$

where $\varepsilon_{AL,T} = \alpha_{AL}\Delta T$ and $\varepsilon_{AL,P} = \frac{\sigma_{AL}}{E_{AL}} = \frac{\frac{Pr}{t_{AL}}}{E_{AL}}$.

Knowing the aluminum properties, $E_{AL} = 10.9 \times 10^6 psi$ and $\alpha_{AL} = 12.8 \times 10^{-6}/°F$, then

$$\varepsilon_{total,AL} = 12.8 \times 10^{-6}(130 - 75) + \frac{-P(2)}{10.9 \times 10^6(0.25)}$$

and the aluminum ring's elongation is

$$\Delta elong_{AL} = 2\pi\frac{4}{2}\left(12.8 \times 10^{-6}(130 - 75) + \frac{-P(2)}{10.9 \times 10^6(0.25)}\right).$$

Thus, equating the steel and aluminum rings' elongations, one would have

$$\Delta elong_{steel} = \Delta elong_{AL}$$

$$2\pi\frac{4}{2}\left(6.5 \times 10^{-6}(130 - 75) + \frac{P(2)}{29 \times 10^6(0.125)}\right)$$

$$= 2\pi\frac{4}{2}\left(12.8 \times 10^{-6}(130 - 75) + \frac{-P(2)}{10.9 \times 10^6(0.25)}\right).$$

Finally, the exerted pressure, P, can be solved for, from the above equality,

$$P = 2737.8 \, psi$$

and the stress on the steel ring can be calculated as

$$\sigma_{steel} = \frac{Pr}{t_{steel}} \text{ or } \sigma_{steel} = \frac{2737.8(2)}{0.125} = 43805 \, psi$$

Example 3.3: (S.I. Units)

A 50.8 mm wide steel cylindrical ring is fitted over a 50.8 mm wide aluminum cylindrical ring, as shown by the following figure. At room temperature, the steel ring has an exact inner diameter of 101.6 mm and is 3.18 mm thick. Also at room temperature, the aluminum ring has an exact outer diameter of 101.6 mm and is 6.35 mm thick. The temperature of the fitting is raised to 54.4°C. Determine the stress on the steel ring and the pressure exerted on the steel ring by the aluminum ring.

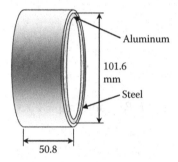

It should be noted that because the two rings are originally fitted exactly over each other and they are at equilibrium at the same temperature, then the steel cylinder's total circumferential elongation is equal to the aluminum cylinder's circumferential elongation after the temperature rise.

Now, the change in the circumferential length of the rings over the original ring circumference is the total strain the ring sees. Thus, the circumferential elongation, $\Delta elong$, can be defined as

$$\Delta elong = 2\pi r \varepsilon_{total}$$

The total strain for the steel cylinder is calculated as the strain due to thermal expansion plus the strain due to the pressure exerted by the aluminum ring expansion,

$$\varepsilon_{total,steel} = \varepsilon_{steel,T} + \varepsilon_{steel,P}$$

where $\varepsilon_{steel,T} = \alpha_{steel}\Delta T$ and $\varepsilon_{steel,P} = \frac{\sigma_{steel}}{E_{steel}} = \frac{\frac{Pr}{t_{steel}}}{E_{steel}}$.

Knowing the steel properties, $E_{steel} = 199948\ MPa$ and $\alpha_{steel} = 12 \times 10^{-6}/°C$, then

$$\varepsilon_{total,steel} = 12 \times 10^{-6}(54.4 - 23.9) + \frac{P(50.8)}{199948(3.18)}$$

and the steel ring's elongation is

$$\Delta elong_{steel} = 2\pi\frac{101.6}{2}\left(12 \times 10^{-6}(54.4 - 23.9) + \frac{P(50.8)}{199948(3.18)}\right).$$

Similarly, for the aluminum ring the total strain is

$$\varepsilon_{total,AL} = \varepsilon_{AL,T} + \varepsilon_{AL,P}$$

where $\varepsilon_{AL,T} = \alpha_{AL}\Delta T$ and $\varepsilon_{AL,P} = \frac{\sigma_{AL}}{E_{AL}} = \frac{\frac{Pr}{t_{AL}}}{E_{AL}}$.

Knowing the aluminum properties, $E_{AL} = 75153\ MPa$ and $\alpha_{AL} = 23 \times 10^{-6}/°C$, then

$$\varepsilon_{total,AL} = 23 \times 10^{-6}(54.4 - 23.9) + \frac{-P(50.8)}{75153(6.35)}$$

and the aluminum ring's elongation is

$$\Delta elong_{AL} = 2\pi\frac{101.6}{2}\left(23 \times 10^{-6}(54.4 - 23.9) + \frac{-P(50.8)}{75153(6.35)}\right).$$

Thus, equating the steel and aluminum rings' elongations, one would have

$$\Delta elong_{steel} = \Delta elong_{AL}$$

$$2\pi\frac{101.6}{2}\left(12 \times 10^{-6}(54.4 - 23.9) + \frac{P(50.8)}{199948(3.18)}\right)$$

$$= 2\pi\frac{101.6}{2}\left(23 \times 10^{-6}(54.4 - 23.9) + \frac{-P(50.8)}{75153(6.35)}\right)$$

Finally, the exerted pressure, P, can be solved for, from the above equality,

$$P = 18 \, MPa$$

and the stress on the steel ring can be calculated as

$$\sigma_{steel} = \frac{Pr}{t_{steel}} \text{ or } \sigma_{steel} = \frac{18(50.8)}{3.18} = 300 \, MPa$$

3.5 THICK-WALLED CYLINDER

For thick-walled cylinders under pressure, the stresses depend on the internal and external pressure and the inner and outer radius of the cylinder. The stress components would be the stresses in the radial direction σ_r and stress in the tangential direction. Figure 3.9 below illustrates the stress components that would exist in a thick-walled cylinder.

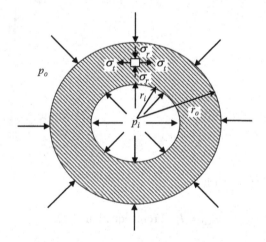

FIGURE 3.9 Thick-walled cylinders under pressure.

The radial stress component is defined by

$$\sigma_r = \frac{p_o r_o^2 - p_i r_i^2 + \left(\frac{r_i^2 r_o^2}{r^2}\right)(p_i - p_o)}{r_o^2 - r_i^2} \tag{3.25}$$

and the tangential stress component is defined by

$$\sigma_t = \frac{p_i r_i^2 - p_o r_o^2 + \left(\frac{r_i^2 r_o^2}{r^2}\right)(p_i - p_o)}{r_o^2 - r_i^2} \tag{3.26}$$

At the radial location $r = r_i$ the maximum stresses occur. By substituting the r_i for r in the equations (3.25) and (3.26) the maximum values are determined.

$$\sigma_r = p_i \tag{3.27}$$

$$\sigma_t = \frac{p_i(r_i^2 + r_o^2) - 2p_o r_o^2}{r_o^2 - r_i^2} \tag{3.28}$$

3.6 STRESS ON HYDRAULIC ACTUATORS

For an actuator under an applied load of F and internal pressure of Pi with an outside diameter of do(=2 ro) and inside diameter of di(=2 ri), one can determine the maximum tangential and radial stress values on the actuator wall are as follows (Figure 3.10):

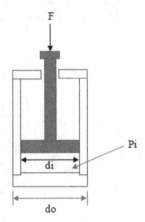

FIGURE 3.10 Hydraulic actuator.

$$\sigma_{rmax} = P_i \ \text{(From equation 3.27)} \tag{3.29}$$

where $Pi = \frac{F}{Ai}$ and Ai is the internal surface area of the actuator head cross-section ($Ai = \pi ri^2$).

Internal area can be written as

$$Ai = \frac{F}{Pi} \tag{3.30}$$

The maximum tangential stress derived from equation (3.28), where external pressure, Po = 0, is

$$\sigma_{tmax} = P_i\left(\frac{ro^2 + ri^2}{ro^2 - ri^2}\right) \tag{3.31}$$

3.7 STRESS DUE TO VARIOUS COMBINED LOADINGS

The stresses on a component typically vary from normal to shear stresses and could also be present in the form of combined states. The best way to categorize them is classification by applied loading type. Refer to Table 3.1 below for the general classification of stresses on a component.

TABLE 3.1

Classification of Stresses on a Component by Various Loading Conditions

Loading Type	Normal Stress	Shear Stress
Axial	$\sigma = \frac{P}{A}$	–
Thermal	$\sigma_T = E\alpha(\Delta T)$	–
Shear	–	$\tau = \frac{V}{A}$
Torsion	–	$\tau = \frac{Tr}{J}$
Bending	$\sigma = \frac{Mc}{I}$	$\tau = \frac{VQ}{Ib}$

Note: P is axial load, V is shear load, A is the cross-sectional area, E is elastic modulus, α is the coefficient of thermal expansion, ΔT is the change in temperature, r is the radius, b is the width, c, is the distance to neutral axis, I is the moment of inertia, J is the polar moment of inertia and Q is the first moment of area in a beam cross-section.

State of stress on a stress element can also be defined as follows per loading condition:

Uniaxial Stress Element:

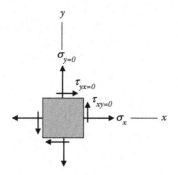

Biaxial Stress Element:

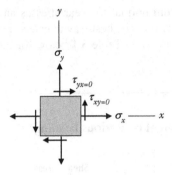

Pure Shear State of Element:

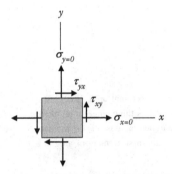

Axial and Bending State of Element:
(Simply-supported beam with concentrated vertical load and applied side loads)

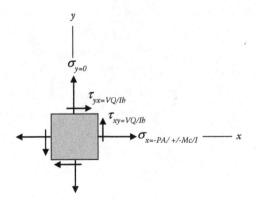

Axial and Torsion State of Element:
(A shaft with torsional load and applied side loads)

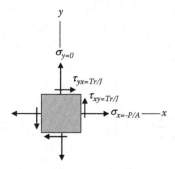

3.8 STRESSES DUE TO ROTATIONAL LOADINGS

Stresses due to rotational loading of a thin disk while spinning are examined here. Assume the following thin disk with the specified parameters in Figure 3.11 that has an internal central opening, (Thin-disk $r_o = 25$ t) t = thickness.

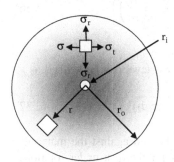

FIGURE 3.11 Rotating thin-disk under loading.

The state of stresses in the rotating thin-disk can be defined by following equation,

$$\sigma_t = \sigma_o \frac{3+v}{8}\left[1 + \left(\frac{ri}{ro}\right)^2 + \left(\frac{ri}{r}\right)^2 - \frac{1+3v}{3+v}\left(\frac{r}{ro}\right)^2\right] \tag{3.32}$$

$$\sigma_r = \sigma_o \frac{3+v}{8}\left[1 + \left(\frac{ri}{ro}\right)^2 - \left(\frac{ri}{r}\right)^2 - \left(\frac{r}{ro}\right)^2\right] \tag{3.33}$$

$$\sigma_o = \rho\,(\omega)^2\,(ro)^2$$

Where

    ρ is the density, lbm/in^3

    ω is the angular velocity, radians/sec.

    r_o is the outside radius of the disk, in.

    r_i is the inside radius of the disk, in.

    and r is the stress element location, in.

Now, σ_{tmax} is at r = ri

$$\sigma_{tmax} = \sigma_o \frac{3+v}{4}\left[1 + \frac{1-v}{3+v}\left(\frac{ri}{ro}\right)^2\right] \tag{3.34}$$

Principal stresses are defined as $\sigma_1 = \sigma_{tmax}$ and $\sigma_2 = 0$

    Also, for σ_{rmax}

$$\sigma_{rmax} = \sigma_o \frac{3+v}{8}\left[1 + \left(\frac{ri}{ro}\right)\right]^2 \tag{3.35}$$

$$\sigma_t = \sigma_o \frac{3+v}{8}\left[1 + \left(\frac{ri}{ro}\right)^2 + 2\frac{1-v}{3+v}\left(\frac{ri}{ro}\right)\right] \tag{3.36}$$

Principal stresses are defined as and $\sigma_1 = \sigma_t$ and $\sigma_2 = \sigma_{rmax}$

    In both of these stress cases the Von-Mises stress level can be defined as

$$\sigma_{von-mises} = \sqrt{(\sigma_1)^2 + (\sigma_2)^2 - (\sigma_1\sigma_2)} \tag{3.37}$$

At this point the Von-Mises stress level can be checked against the material's yield strength to determine the margins of safety for the disk under rotational loading.

Problems

1. A cylindrical vessel with both ends closed, has a wall thickness of 0.08 in., a diameter of 10 in. and a depth of 25 in. long. The vessel is internally pressurized to 1000 psi. Determine the hoop and long-itudinal stresses on the vessel walls if the vessel is made of steel ASTM A514.

2. A compound vessel with an internal diameter of 3 in. is made out of 0.2 in. thick steel and 0.15 in. thick copper for external and internal layers, respectively. The vessel is 20 in. long. The vessel's temperature is raised 20°F. What is the contact pressure generated between the two layers of the vessel?

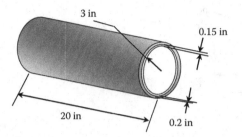

3. For the steel vessel shown below, the physical dimensions are, h = 90 ft, r = 9 ft. For this vessel with an allowable internal pressure of 20,000 psi, determine the vessel's required thickness.

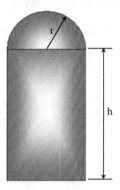

4. An aluminum 7075 thick-walled cylinder of 2 in. thick with an internal radius of 1 in. is pressurized internally to 1000 psi. What are the tangential and radial stresses on the cylinder wall?

5. A ball bearing made out of steel balls with 0.3 in. in diameter is fitted over a shaft. The shaft is exerting an radial force of 100 lbf on the bearing. If the bearing ring container is also made out of steel, what is the contact stress on the balls?

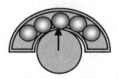

6. A pipe with internal radius of r is pressurized by a pressure of Pi and installed by bolts to a fixed space, as shown by the following figure. Determine the state of combined stresses on the pipe. Hint: σ_{axial} due to installation is equal to $\sigma_{axial} = (E)(\Delta L/L)$

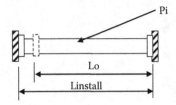

REFERENCES

Beer, F.P., Johnston, E.R., DeWolf, J.T., *Mechanics of Material*, 2002. New York: MacGraw Hill.

Shigley, J.E., Mischke, C.R., *Mechanical Engineering Design*, 1989. New York: McGraw Hill.

Ugural, A.C., Fenster, S.K., *Advanced Strength and Applied Elasticity*, 1995. New Jersey: Prentice Hall.

Young, W.C., *Roark's Formulas for Stress & Strain*, 1989. New York: McGraw Hill.

4 Failure Criteria

4.1 INTRODUCTION

This chapter covers theory of failure criteria for ductile and brittle materials. The different criteria known are defined, and their application is elaborated. It is assumed that the reader is familiar with the concept of the principal stresses since the failure criteria for metallic components are mainly based on principal stresses. In addition, the topics of the factor of safety and the calculations of stress margins of safety are briefly discussed in this chapter.

4.2 MATERIAL FAILURE

Failure is defined as the inability of a structural component to withhold applied loading. Failure can be based on one or several factors, but normally in engineering it is based on stress, strain, deflection, crack length or number of residual life cycles. In metals, failure criteria are typically based on stress, and in composites, failure criteria are based on the strain. Also, in general, structural metal behavior is defined as either ductile or brittle.

4.2.1 DUCTILE FAILURE

For metallic structural components that are made from ductile materials, which have elongation capability of more than 5%, the yielding is the criteria for failing. In this type of material failure, often the compression and tensile yielding are at the same strength ($\sigma_{yld} = \sigma_{yldt} = -\sigma_{yldc}$). There are two known yield failure criteria: the maximum-shearing-stress criterion and the maximum-distortion-energy criterion (Von-Mises). Thus, the general accepted failure theories under the ductile materials are:

1. Maximum Shearing-Stress Theory
2. Maximum Distortion Energy Theory

4.2.2 BRITTLE FAILURE

For components made from brittle materials, which have elongation capability of less than 5%, the fracture is the criteria for failing. In this type of material failure, often the ultimate compression and ultimate tensile are the failure limits. Also, there are two known fracture failure criteria: the maximum-normal-stress criterion and the Coulomb-Mohr's criterion. Thus, the general accepted failure theories under the brittle materials are:

DOI: 10.1201/9781003311218-4

1. Maximum Normal Stress Theory
2. Coulomb-Mohr Theory

4.3 MAXIMUM-SHEARING-STRESS CRITERION (DUCTILE MATERIAL)

The maximum shearing stress criterion states that the material has failed when the shearing stress on the component has reached the yield shear strength of the material, which is derived from the tensile test of the material specimen. The yield shear strength is defined as the one-half of the tensile yield strength (σ_{yld}) of the material.

However, structural components are normally under multi-axial loading. That is, in 2-D plane stress condition the maximum shear stress of the component is equal to one-half of the maximum normal stress or equal to one-half of the difference between the maximum and minimum normal stress if the maximum stress is tensile and the minimum stress is compressive.

In more general terms, this criterion states that if principal stresses of the component have the same signs, the magnitudes of the principal stresses have to be less than the yield strength of the material. Similarly, if the principal stresses have different signs, then the difference between the principal stresses has to be less than the material yield strength.

Thus, for same signs:

$$|\sigma_1| < \sigma_{yld} \text{ and } |\sigma_2| < \sigma_{yld} \tag{4.1}$$

For different signs:

$$|\sigma_1 - \sigma_2| < \sigma_{yld} \tag{4.2}$$

For justification of equation (4.2), it should be reminded that, $\tau_{max} = |\sigma_1 - \sigma_2| < \sigma_{yld}$ $\tau_{max} = (1/2)(\sigma_1 - \sigma_2)$ and $\tau_{yld} = (1/2)\,\sigma_{yld}$; thus, in comparison, the (1/2) factor can be eliminated, and equation (4.2) is readily developed.

This is graphically shown by Figure 4.1 below. The state of stress outside of the shaded area is considered to be in a failure state.

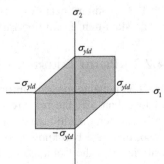

FIGURE 4.1   The maximum-shearing-stress criterion.

4.4 MAXIMUM-DISTORTION-ENERGY CRITERION (DUCTILE MATERIAL)

The maximum-distortion-energy criterion indicates that the material failure occurs when the distortion energy of a component reaches the energy for yielding. That is, for a component to be "safe", the principal stress relation below has to be met:

$$\sigma_1^2 - \sigma_1\sigma_2 + \sigma_2^2 < \sigma_{yld}^2 \qquad (4.3)$$

Graphically, this Von-Mises criterion is shown by Figure 4.2 below. Once again for the structural component to be "safe", the state of stress has to fall within the shaded region shown.

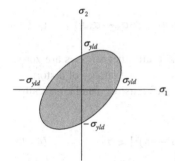

FIGURE 4.2   The maximum-distortion-energy criterion.

In this failure theory, when at any point in the element the distortion energy per unit volume equals the yielding in a simple tension test, failure occurs. Strain energy, U, is given by:

$$U = (1/2)\sigma\varepsilon \qquad (4.4)$$

where in a combined state of stress, the strain components are given as:

$$
\begin{aligned}
\varepsilon 1 &= (1/E)(\sigma_1 - v\sigma_2 - v\sigma_3) \\
\varepsilon 2 &= (1/E)(-v\sigma_1 + \sigma_2 - v\sigma_3) \\
\varepsilon 3 &= (1/E)(-v\sigma_1 - v\sigma_2 + \sigma_3)
\end{aligned}
\qquad (4.5)
$$

Substituting (4.5) in the strain energy equation above (4.4), one would have:

$$U = (1/2)\left[\begin{array}{l} \sigma 1(1/E)(\sigma 1 - v\sigma 2 - v\sigma 3) + \sigma 2(1/E)(-v\sigma 1 + \sigma 2 - v\sigma 3) \\ + \sigma 3(1/E)(-v\sigma 1 - v\sigma 2 + \sigma 3) \end{array}\right]$$

$$(4.6)$$

By substituting the average stress,

$$\sigma_{av} = (\sigma_1 + \sigma_2 + \sigma_3)(1/3) \qquad (4.7)$$

into the equation (4.6) above, one would have:

$$U_v = (3/2E)\sigma_{av^2}(1 - 2\nu) \qquad (4.8)$$

Then, substitute equation (4.7) into this equation, and the results is:

$$U_v = ((1 - 2\nu)/6E)(\sigma_1^2 + \sigma_2^2 + \sigma_3^2 + 2\sigma_1\sigma_2 + 2\sigma_2\sigma_3 + 2\sigma_3\sigma_1) \qquad (4.9)$$

Knowing that the distortion energy is $U_d = U - U_v$, then

$$U_d = ((1 + \nu)/(3E))(\sigma_1^2 + \sigma_2^2 + \sigma_3^2 - \sigma_1\sigma_2 - \sigma_2\sigma_3 - \sigma_3\sigma_1) \qquad (4.10)$$

For a simple tensile test at yielding the $\sigma_1 = \sigma_{yld}$ and all other stresses are zero. Thus, Von Mises criterion states that failure occurs when the energy of distortion reaches the same energy for yield/failure in uniaxial tension. Mathematically, this is expressed as

$$(1/2)[(\sigma_1 - \sigma_2)^2 + (\sigma_2 - \sigma_3)^2 + (\sigma_3 - \sigma_1)^2] < \sigma_{yld}^2 \qquad (4.11)$$

In the cases of plane stress, $\sigma_3 = 0$. The Von Mises criterion reduces to

$$\sigma_1^2 - \sigma_1\sigma_2 + \sigma_2^2 < \sigma_{yld}^2 \qquad (4.12)$$

4.5 MAXIMUM-NORMAL-STRESS CRITERION (BRITTLE MATERIAL)

The maximum-normal-stress criterion for brittle material states that the component failure occurs when the principal stresses have reached the ultimate tensile or compressive strength of the material derived from the tensile test of the material specimen. That is, for the component to be "safe" the state of stress has to be less than the ultimate strength, as shown by Figure 4.3 below. Thus,

$$|\sigma_1| < \sigma_{ult} \text{ and } |\sigma_2| < \sigma_{ult} \qquad (4.13)$$

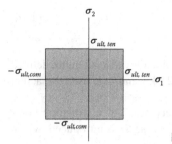

FIGURE 4.3   The maximum-normal-stress criterion.

4.6   COULOMB-MOHR'S CRITERION (BRITTLE MATERIAL)

The Coulomb-Mohr's criterion states that the component is "safe" if when both principal stresses are positive, they are less than the ultimate tensile strength.

$$\sigma_1 < \sigma_{ult,ten} \text{ and } \sigma_2 < \sigma_{ult,ten} \tag{4.14}$$

Similarly, it states the component is "safe" if when both principal stresses are negative, their magnitudes are less than the magnitude of the ultimate compressive strength.

$$|\sigma_1| < |\sigma_{ult,com}| \text{ and } |\sigma_2| < |\sigma_{ult,com}| \tag{4.15}$$

When they have opposite signs, they should fall within the shaded region shown by Figure 4.4 below.

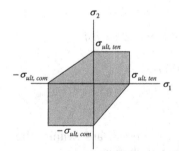

FIGURE 4.4   The Coulomb-Mohr's criterion.

4.7   FACTOR OF SAFETY CALCULATIONS

The factor of safety (F.S.) is normally referred to as the ratio of the ultimate (allowable) strength over the actual stress the component undergoes.

$$F.S. = \frac{\sigma_{allowable}}{\sigma_{actual}} \tag{4.16}$$

The factor of safety based on the failure criteria is the allowable strength, defined by the shaded regions shown by Figures 4.1, 4.2, 4.3 and 4.4, over the actual state of stress represented by the principal stresses.

Now to determine the percentage of the "safety-ness" of a structural component under loading, the concept of margin of safety (*M.S.*) is introduced, which simply is the following ratio:

$$M.S. = \frac{\sigma_{allowable}}{(F.S.)\sigma_{actual}} - 1 \tag{4.17}$$

In design of structures this ratio is calculated to determine the adequacy of the design based on a predetermined factor of safety. Usually, if the allowable used is the yield strength of the material, a factor of $F.S. = 1.5$ is used for margin of safety calculations. Likewise, if the allowable used is the ultimate strength of the material, a factor of $F.S. = 2.0$ is used for margin of safety calculations.

Example 4.1: Reconsider Example 2.2 shown in chapter 2; assume it is made out of aluminum 6061-T6 (σ_{yld} = 35000 psi). Using the maximum-shearing-stress criterion, determine the factor of safety for the component. Also, by assuming a factor of safety of 1.0, determine the margin of the safety for this component.

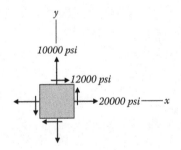

The principal stresses are:

$$\sigma_{1,2} = = 28000 \text{ and } 2000 \text{ psi}$$

$$\tau_{1,2} = \pm 13000 \text{ psi}$$

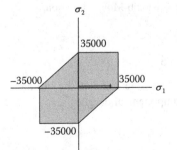

The principal stresses are within the shaded region as shown:

The factor of safety is:

$$F.S. = \frac{(1/2)\sigma_{yld}}{\tau} = \frac{(1/2)(35000)}{13000} = 1.35$$

Note: Shear allowable is ½ of the material yield strength.
The margin of safety with use of a 1.0 factor of safety and a maximum actual stress of 28,000 psi is calculated as

$$M.S. = \frac{\sigma_{allowable}}{(F.S.)\sigma_{actual}} - 1$$

$$M.S. = \frac{35000}{(1)(28000)} - 1 = 0.25 \text{ or } 25\%$$

Example 4.1: (S.I. Units)

Reconsider Example 2.2 shown in chapter 2; assume it is made out of aluminum 6061-T6 ($\sigma_{yld} = 241.3 MPa$). Using the maximum-shearing-stress criterion, determine the factor of safety for the component. Also, by assuming a factor of safety of 1.0, determine the margin of the safety for this component.

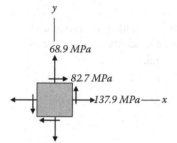

The principal stresses are:

$\sigma_{1,2} = = 193.1 \ MPa \ and \ 13.8 \ MPa$

$\tau_{1,2} = \pm 89.6 \ MPa$

The principal stresses are within the shaded region as shown:

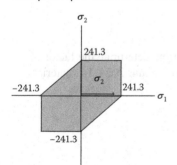

The factor of safety is:

$$F.S. = \frac{(\frac{1}{2})\sigma_{yld}}{\tau} = \frac{(\frac{1}{2})(241.3)}{89.6} = 1.35$$

Note: Shear allowable is ½ of the material yield strength.

The margin of safety with use of a 1.0 factor of safety and a maximum actual stress of 28,000 psi is calculated as

$$M.S. = \frac{\sigma_{allowable}}{(F.S.)\sigma_{actual}} - 1$$

$$M.S. = \frac{241.3}{(1)(193.1)} - 1 = 0.25 \text{ or } 25\%$$

Problems

1. Using the maximum-shearing stress criterion and the maximum-distortion-energy criterion, determine the factor of safety for the state of stress shown for a steel ASTM-A36 material.

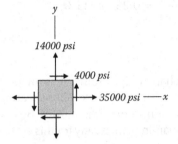

2. Using the maximum-shearing stress criterion and the maximum-distortion-energy criterion, determine the factor of safety for the state of stress shown for an aluminum 6061-T6 material.

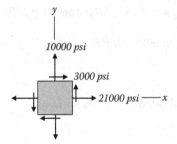

3. Using the maximum-distortion-energy criterion, determine the factor of safety for the state of stress shown for an aluminum 2024-T4 material.

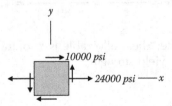

4. Using the Mohr's criterion, determine if the material would fail for the state of stress shown for a steel 6061-T6 material.

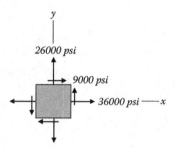

5. The maximum stress seen by a component made out of steel ASTM-A242 material is 43,000 psi. Determine the margin of safety for yielding for this material.

REFERENCES

Beer, F.P., Johnston, E.R., DeWolf, J.T., *Mechanics of Material*, 2002. New York: McGraw Hill.

Shigley, J.E., Mischke, C.R., *Mechanical Engineering Design*, 1989. New York: McGraw Hill.

Ugural, A.C., Fenster, S.K., *Advanced Strength and Applied Elasticity*, 1995. New Jersey: Prentice Hall.

5 Beam Analysis Theory

5.1 INTRODUCTION

This chapter covers the theory behind the analysis of the beams under various types of loading. The shear and moment diagram construction is elaborated. The deflection analysis of the beam based on bending curvature theory, by double integration method, is discussed. The bending, shear and torsion of the beams for determination of the stresses are examined here. Although beam analysis can be a lengthy topic, only the essentials that are needed to fully perform a stress analysis are presented here.

5.2 BOUNDARY CONDITIONS AND LOADING

Before analyzing the beams for either deflection or stress, it is essential to understand the loading and the boundary conditions that typically arise from the design of structural systems with beams. The typical classifications of the beam structural systems based on boundary conditions are: simply supported beam, fixed beam, one side fixed and one side simply supported beam and cantilevered beam. Figure 5.1 below illustrates these different boundary conditions.

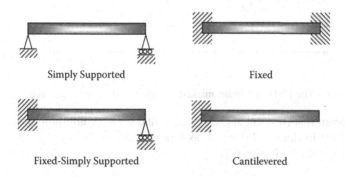

Simply Supported

Fixed

Fixed-Simply Supported

Cantilevered

FIGURE 5.1 The typical beam boundary conditions.

The loading on the beams can be classified as: concentrated loading, uniformly distributed loading, gradual increasing or decreasing loading, moment or couple loading and torsion loading. Concentrated loading, uniformly distributed loading and gradual loading can produce both bending and shear of the beam. Moment or couple loading can produce bending on the beam. Torsion loading can produce twisting on the beam.

DOI: 10.1201/9781003311218-5

5.3 SHEAR AND MOMENT DIAGRAMS

With the aim of determining the bending and shear stresses on any beam structure, the shear and moment diagrams have to be constructed. The shear and bending moment on a beam at any location along the length of the beam can be calculated by setting the beam at static equilibrium, where

$$\sum F = 0 \tag{5.1}$$

and

$$\sum M = 0 \tag{5.2}$$

To determine the shear, $V(x)$, and bending, $M(x)$, loading at any point along the length of the beam, simply the state of the beam has to be analyzed after each point of the application of the loading. Figure 5.2 illustrates an example of a beam free-body diagram (FBD) under loading with the locations that require the beam to be analyzed at, marked by numbers 1 through 4.

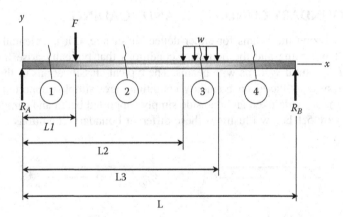

FIGURE 5.2 The FBD of a beam marked for shear and moment analysis.

The beam is section cut at each location and set to equilibrium for both shear and moment loading at distance x, as follows,
  For $0 < x < L1$: (Figure 5.3)

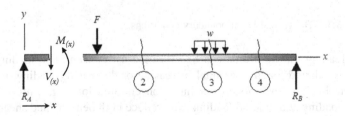

FIGURE 5.3 The FBD of the beam sectioned at point (1).

$$+ up \sum F_y = 0 \qquad R_A - V_{(x)} = 0 \qquad V_{(x)} = R_A \qquad (5.3)$$

$$+ ccw \sum M = 0 \qquad - R_A x + M_{(x)} = 0 \qquad M_{(x)} = R_A x \qquad (5.4)$$

For $L1 < x < L2$: (Figure 5.4)

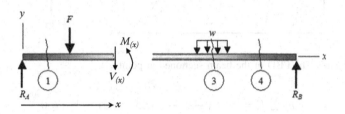

FIGURE 5.4 The FBD of the beam sectioned at point (2).

$$+ up \sum F_y = 0 \qquad R_A - F - V_{(x)} = 0 \qquad V_{(x)} = R_A - F \qquad (5.5)$$

$$+ ccw \sum M = 0 \qquad - R_A x + F(x - L1) + M_{(x)} = 0 \qquad M_{(x)} = R_A x - F(x - L1)$$
$$(5.6)$$

For $L2 < x < L3$: (Figure 5.5)

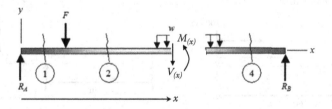

FIGURE 5.5 The FBD of the beam sectioned at point (3).

$$+ up \sum F_y = 0 \qquad R_A - F - w(x - L2) - V_{(x)} = 0$$
$$V_{(x)} = R_A - F - w(x - L2) \qquad (5.7)$$

$$+ ccw \sum M = 0$$
$$- R_A x + F(x - L1) + w(x - L2)\left(\frac{x - L2}{2}\right) + M_{(x)} = 0$$
$$M_{(x)} = R_A x - F(x - L1) - w\frac{(x - L2)^2}{2} \qquad (5.8)$$

For $L3 < x < L$: (Figure 5.6)

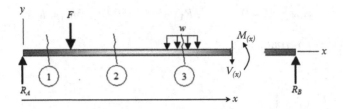

FIGURE 5.6 The FBD of the beam sectioned at point (4).

$$+ up \, \Sigma \, F_y = 0 \quad R_A - F - w(L3 - L2) - V_{(x)} = 0$$

$$V_{(x)} = R_A - F - w(L3 - L2) \tag{5.9}$$

$$+ ccw \, \Sigma \, M = 0$$

$$- R_A x + F(x - L1) + w(L3 - L2)\left(x - \frac{L3 + L2}{2}\right) + M_{(x)} = 0 \tag{5.10}$$

$$M_{(x)} = R_A x - F(x - L1) - w(L3 - L2)\left(x - \frac{L3 + L2}{2}\right)$$

Example 5.1: For the beam system shown below, determine the shear-moment diagrams.

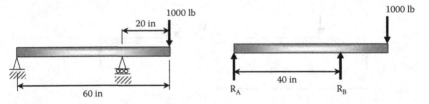

From the FBD, the reactions are calculated as

$$R_A = -500 \text{ lb} \quad \text{and} \quad R_B = 1500 \text{ lb}$$

For $0 < x < 40$:

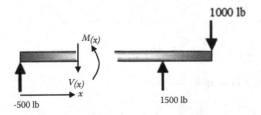

$$+ up \sum F_y = 0, \quad -500 - V_{(x)} = 0 \quad V_{(x)} = -500 \text{ lbf}$$

$$+ ccw \sum M = 0, \quad -(-500)x + M_{(x)} = 0 \quad M_{(x)} = -500x \text{ in} - \text{lb}$$

For $40 < x < 60$:

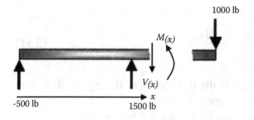

$$+ up \sum F_y = 0, \quad -500 + 1500 - V_{(x)} = 0 \quad V_{(x)} = 1000 \text{ lbf}$$

$$+ ccw \sum M = 0, \quad -(-500)x - 1500(x - 40) + M_{(x)} = 0$$
$$M_{(x)} = -500x + 1500(x - 40) \text{ in} - \text{lb}$$

Based on this FBD,

The shear and moment diagram are

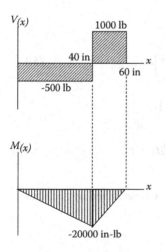

5.4   DEFLECTION OF THE BEAMS

Here the deflection of the beam is derived from the curvature representation theory of the beam. Basically the curvature differential equation of the beam is integrated twice, to first produce the slope of the beam and then to produce the deflection of the beam.

The equation of the curvature of the beam learned from the mechanics of solid is

$$\frac{d^2v}{dx^2} = -\frac{M(x)}{EI} \tag{5.11}$$

where $M(x)$ is the bending moment, v is the displacement in vertical direction, E is the elastic modulus and I is the moment of inertia.

The orientation of the beam is defined by Figure 5.7 below.

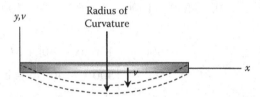

FIGURE 5.7   The beam deflection orientation.

Simply by separation of the variables and double integration the deflection is derived from expression (5.11):

$$EI \iint d^2v = \int_0^x \int_0^x -M(x)dx^2 \tag{5.12}$$

Integrating once, one would have (the slope)

$$EIdv = \left(\int_0^x -M(x)dx + C1 \right)dx \tag{5.13}$$

and, by integrating the second time, the deflection is

$$EIv = \int_0^x \left(\int_0^x -M(x)dx + C1 \right)dx + C2 \tag{5.14}$$

where the constants of the integration, $C1$ and $C2$, are determined by application of the appropriate boundary conditions.

Example 5.2: Consider a simply supported beam of length L under a concentrated loading, P, at a location, a, distance away from the edge of the beam. Determine the deflection of the beam along the length (x) in y-direction.

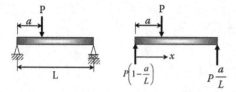

Let v_1, represent the deflection for segment $x \leq a$, and v_2 represent the deflection for segment $x \geq a$.

The applicable boundary conditions are:

$$v_1 = 0 \text{ at } x = 0$$

$$v_2 = 0 \text{ at } x = L$$

$$v_1 = v_2 \text{ and } \frac{dv_1}{dx} = \frac{dv_2}{dx} \text{ at } x = a$$

For $x \leq a$:

Knowing that the bending moment on the beam is

$M(x) = P\left(1 - \frac{a}{L}\right)x$, by expression (5.11) the beam curvature is

$EI\frac{d^2v_1}{dx^2} = -M(x) = -P\left(1 - \frac{a}{L}\right)x$; integrating once, one would have

$EI\frac{dv_1}{dx} = \frac{-Px^2}{2} + \frac{Pax^2}{2L} + C1$; integrating a second time, one would have

$EIv_1 = \frac{-Px^3}{6} + \frac{Pax^3}{6L} + C1x + C2$ for the deflection.

For $x \geq a$:

Knowing that the bending moment on the beam for this segment of the beam is

$M(x) = P\left(1 - \frac{a}{L}\right)x - P(x - a)$, by expression (5.11) the beam curvature is

$EI\frac{d^2v_2}{dx^2} = -M(x) = -P\left(1 - \frac{a}{L}\right)x + P(x - a)$; integrating once, one would have

$EI\frac{dv_2}{dx} = -P\left(1 - \frac{a}{L}\right)\frac{x^2}{2} + \frac{Px^2}{2} - Pax + C3$; integrating a second time, one would have

$EIv_2 = -P\left(1 - \frac{a}{L}\right)\frac{x^3}{6} + \frac{Px^3}{6} - \frac{Pax^2}{2} + C3x + C4$ for the deflection.

Applying the boundary conditions and solving for the integration constants, one would have

$$C1 = \frac{Pa}{6L}(a - 2L)(a - L)$$

$$C2 = 0$$

$$C3 = \frac{Pa}{6L}(a^2 + 2L^2)$$

$$C4 = -\frac{Pa^3}{6}$$

Thus, the deflection of the beam under a concentrated loading at an intermediate location is

For $x \le a$:

$$v_1 = \frac{1}{EI}\left(-\frac{Px^3}{6} + \frac{Pax^3}{6L} + \frac{Pa}{6L}(a - 2L)(a - L)x\right)$$

For $x \ge a$:

$$v_2 = \frac{1}{EI}\left(\frac{Pax^3}{6L} - \frac{Pax^2}{2} + \frac{Pa}{6L}(a^2 + 2L^2)x - \frac{Pa^3}{6}\right)$$

Example 5.3: Consider the cantilevered beam of length L, shown below under an uniformly distributed loading of w. For this beam, determine the deflection of the beam in vertical y-direction along the length of the beam (x).

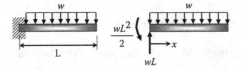

The applicable boundary conditions are:

$$v = 0 \text{ and } \frac{dv}{dx} = 0 \text{ at } x = 0$$

Knowing that the bending moment on the beam is

$M(x) = -\frac{wx^2}{2} + wLx - \frac{wL^2}{2}$, by expression (5.11) the beam curvature is

$EI\frac{d^2v}{dx^2} = -M(x) = \frac{wx^2}{2} - wLx + \frac{wL^2}{2}$; integrating once, one would have

$EI\frac{dv}{dx} = \frac{wx^3}{6} - \frac{wLx^2}{2} + \frac{wL^2x}{2} + C1$; integrating a second time, one would have

$EIv = \frac{wx^4}{24} - \frac{wLx^3}{6} + \frac{wL^2x^2}{4} + C1x + C2$ for the deflection.

Applying the boundary conditions and solving for the integration constants, one would have

$$C1 = 0$$

$$C2 = 0$$

Thus, the deflection of the beam under a uniformly distributed loading is

$$v = \frac{1}{EI}\left(\frac{wx^4}{24} - \frac{wLx^3}{6} + \frac{wL^2x^2}{4}\right)$$

Example 5.4: Consider the cantilevered beam of length L, shown below under a gradually increasing distributed loading of w_0. For this beam, determine the deflection of the beam in vertical y-direction along the length of the beam (x).

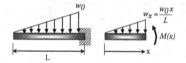

The applicable boundary conditions are:

$$v = 0 \text{ and } \frac{dv}{dx} = 0 \text{ at } x = L$$

Knowing that the bending moment on the beam is

$M(x) + \left(\frac{1}{2}\frac{w_0x^2}{L}\right)\frac{x}{3} = 0$ or $M(x) = -\frac{1}{6}\frac{w_0x^3}{L}$, by expression (5.11) beam curvature is

$EI\frac{d^2v}{dx^2} = -M(x) = \frac{1}{6}\frac{w_0x^3}{L}$; integrating once, one would have

$EI\frac{dv}{dx} = \frac{1}{24}\frac{w_0x^4}{L} + C1$; integrating a second time, one would have

$EIv = \frac{1}{120}\frac{w_0x^5}{L} + C1x + C2$, for the deflection.

Applying the boundary conditions and solving for the integration constants, one would have

$$C1 = -\frac{w_0 L^3}{24}$$

$$C2 = \frac{w_0 L^4}{30}$$

Thus, the deflection of the beam under a uniformly distributed loading is

$$v = \frac{1}{EI}\left(\frac{1}{120}\frac{w_0 x^5}{L} - \frac{w_0 L^3 x}{24} + \frac{w_0 L^4}{30}\right)$$

Example 5.5: Consider the cantilevered beam of length L, shown below under an uniformly distributed loading of w. The beam is simply supported at the right-hand side, as shown. For this beam, determine the deflection of the beam in vertical y-direction along the length of the beam (x-direction).

By principals of superposition this structural system is equivalent to the following two systems:

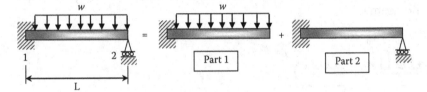

Notice part one of this equality is solved by Example 5.3 already, and the deflection of the beam was determined to be

$$v = \frac{1}{EI}\left(\frac{wx^4}{24} - \frac{wLx^3}{6} + \frac{wL^2 x^2}{4}\right)$$

Then, part two of this equality is represented as follows:

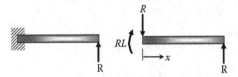

The moment on this beam along the x-direction is

$M(x) = RL - Rx$; by expression (5.11) the beam curvature is

$EI\frac{d^2 v}{dx^2} = -M(x) = -RL + Rx$; integrating once, one would have,

$EI\frac{dv}{dx} = -RLx + \frac{Rx^2}{2} + C1$; integrating a second time, one would have

$EIv = -\frac{RLx^2}{2} + \frac{Rx^3}{6} + C1x + C2$ for the deflection.

Applying the boundary conditions,

$$v = 0 \text{ and } \frac{dv}{dx} = 0 \text{ at } x = 0$$

and solving for the integration constants, one would have

$$C1 = 0$$

$$C2 = 0$$

Thus, the deflection of the beam under a uniformly distributed loading is

$$v = \frac{1}{EI}\left(\frac{Rx^3}{6} - \frac{RLx^2}{2}\right)$$

Now, combining the deflection expressions from part one and part two, one would have the following expression for the total deflection of the original system in question:

$$v = \frac{1}{EI}\left(\frac{wx^4}{24} - \frac{wLx^3}{6} + \frac{wL^2x^2}{4} + \frac{Rx^3}{6} - \frac{RLx^2}{2}\right)$$

To determine the reaction load, R and the moment equation for the beam, one can use the fact that the deflection at point 2 (at $x = L$) is zero and solve for reaction R in the above total deflection equation.

$$v(x = L) = \frac{1}{EI}\left(\frac{wL^4}{24} - \frac{wL^4}{6} + \frac{wL^4}{4} + \frac{RL^3}{6} - \frac{RL^3}{2}\right) = 0,$$

Hence, reaction $R = \frac{3wL}{8}$. Substituting R back into the total deflection expression, one would have

$$v = \frac{1}{EI}\left(\frac{wx^4}{24} - \frac{wLx^3}{6} + \frac{wL^2x^2}{4} + \frac{3wLx^3}{48} - \frac{3wL^2x^2}{16}\right) \text{ or,}$$

$$v = \frac{1}{EI}\left(\frac{wx^4}{24} - \frac{5wLx^3}{48} + \frac{wL^2x^2}{16}\right)$$

Now, to determine the moment on the beam along the x-direction, refer to the free-body diagram of the original system

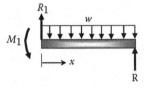

whereas by summation of the moments and forces at point 1, M_1 is determined to be

$$M_1 = \frac{wL^2}{8} \text{ and } R_1 = \frac{5wL}{8}.$$

Thus, at any point along the beam x-direction, the moment for the original beam system is

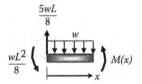

$$M(x) = -\frac{wx^2}{2} + \frac{5wLx}{8} - \frac{wL^2}{8}$$

Example 5.6: Consider a simply supported beam of length L, under a bending moment, M, at a location, a, distance away from the edge of the beam. Determine the deflection of the beam along the length (x) in y-direction.

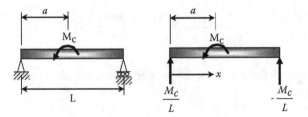

Let v_1 represent the deflection for segment $x \leq a$ and v_2 represent the deflection for segment $x \geq a$.

The applicable boundary conditions are:

$$v_1 = 0 \text{ at } x = 0$$

$$v_2 = 0 \text{ at } x = L$$

$$v_1 = v_2 \text{ and } \frac{dv_1}{dx} = \frac{dv_2}{dx} \text{ at } x = a$$

For $x \leq a$:
Knowing that the bending moment on the beam is

$M(x) = \frac{M_c x}{L}$, by expression (5.11) the beam curvature is

$EI\frac{d^2 v_1}{dx^2} = -M(x) = -\frac{M_c x}{L}$; integrating once, one would have

$EI\frac{dv_1}{dx} = -\frac{M_c x^2}{2L} + C1$; integrating a second time, one would have

$EIv_1 = -\frac{M_c x^3}{6L} + C1x + C2$ for the deflection.

For $x \geq a$:
Knowing that the bending moment on the beam for this segment of the beam is

$M(x) = \frac{M_c x}{L} - M_c$, by expression (5.11) the beam curvature is

$EI\frac{d^2 v_2}{dx^2} = -M(x) = -\frac{M_c x}{L} + M_c$; integrating once, one would have

$EI\frac{dv_2}{dx} = -\frac{M_c x^2}{2L} + M_c x + C3$; integrating a second time, one would have

$EIv_2 = -\frac{M_c x^3}{6L} + \frac{M_c x^2}{2} + C3x + C4$ for the deflection.

Applying the boundary conditions and solving for the integration constants, one would have

$$C1 = -\frac{M_c a^2}{2L} - \frac{M_c L}{3} + M_c a$$

$$C2 = 0$$

$$C3 = -\frac{M_c L}{3} - \frac{M_c a^2}{2L}$$

$$C4 = \frac{M_c a^2}{2}$$

Thus, the deflection of the beam under a concentrated loading at an intermediate location is
For $x \leq a$:

$$v_1 = \frac{M_c}{EI}\left(-\frac{x^3}{6L} + \left(-\frac{a^2}{2L} - \frac{L}{3} + a\right)x\right)$$

For $x \geq a$:

$$v_2 = \frac{M_c}{EI}\left(-\frac{x^3}{6L} + \frac{x^2}{2} - \frac{Lx}{3} - \frac{a^2x}{2L} + \frac{a^2}{2}\right)$$

5.5 BENDING OF THE BEAMS

The moments on the beam cross-section about the in-plane axes z and y cause a bending stress on the beam member, in the normal x-direction, as illustrated by Figure 5.8 below.

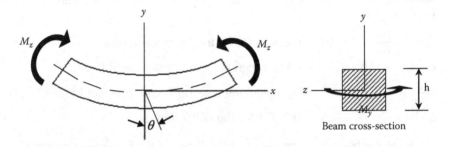

FIGURE 5.8   The bending of the beam about z and y axes.

This normal stress due to the beam bending is defined as

$$\sigma_x = \frac{M_z y}{I_z} + \frac{M_y z}{I_y} \tag{5.15}$$

where M_z and M_y are the moments about the z and y axis, respectively, I_z and I_y are the moments-of-inertia about the z and y axis, respectively, and y, as well as z, are the locations along the beam cross-section from the neural axis to the boundaries of the beam cross-section.

This normal stress may have another component, which is due to any perpendicular axial loading acting on the cross-section of the beam. Hence, expanding the normal stress expression to the following form:

$$\sigma_x = \frac{P_x}{A} + \frac{M_z y}{I_z} + \frac{M_y z}{I_y} \tag{5.16}$$

where P_x is the axial load on the cross-section and A is the area of the cross-section.

Now, it should be noted that the maximum normal stress due to any bending moment (M) occurs at the extreme boundary location at $\frac{h}{2}$ for a symmetric cross-section. Thus, this maximum normal stress is expressed as

$$\sigma_{max} = \frac{Mh}{2I} \qquad (5.17)$$

5.6 SHEAR OF THE BEAMS

Normally the loads acting parallel to the plane of the beam cross-section cause shear stresses on the beam. The shear stress in the beam cross-section is defined as

$$\tau_{xy} = \frac{VQ}{It} \qquad (5.18)$$

where the V is the shear load, Q is the first moment of area, I is the total section moment of inertia and t is the width of the section (Figure 5.9).

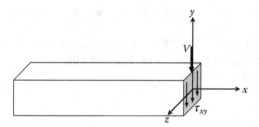

FIGURE 5.9   The shear of the beam cross-section.

The first moment of area Q is a function of the area and the distance from the neural axis (N.A.) to the section where stress needs to be calculated. The Q is defined as

$$Q = \int_A y\, dA \qquad (5.19)$$

or, for combined sections made of several area cross-sections, Q is defined as

$$Q = \sum_i A_i \bar{y}_i = A_1 \bar{y}_1 + A_2 \bar{y}_2 + \dots\dots \qquad (5.20)$$

The $\bar{y}$ and A are evaluated as

$$\bar{y} = \frac{1}{2}(y_t + y) \tag{5.21}$$

$$A = (y_t - y)b \tag{5.22}$$

where y_t is the distance from the N.A. to the top of the section and b is the width of the section y-distance away from the N.A.

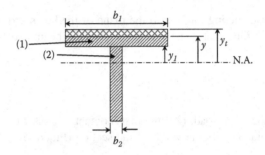

FIGURE 5.10 The beam cross-section first moment of area segment.

As an example, let's use a "T" section or the top part of an "I" section to illustrate the calculation method of Q (Figure 5.10).

The $\bar{y}_1$ and A_1 are evaluated as

$$\bar{y}_1 = \frac{1}{2}(y_t + y) \tag{5.23}$$

$$A_1 = (y_t - y)b_1 \tag{5.24}$$

and the first moment of area is

$$Q_1 = (y_t - y)b_1\frac{1}{2}(y_t + y) \tag{5.25}$$

Thus, by expression (5.18) the shear stress in part (1) is

$$\tau_{xy} = \frac{VQ_1}{Ib_1} \tag{5.26}$$

Similarly, for part (2) of the beam cross-section, the first moment of area is calculated as

$$Q_2 = A_1\bar{y}_1 + A_2\bar{y}_2 = (y_t - y_1)b_1\frac{1}{2}(y_t + y_1) + (y_1 - y)b_2\frac{1}{2}(y_1 + y) \quad (5.27)$$

Thus, again by expression (5.18) the shear stress in part (2) is

$$\tau_{xy} = \frac{VQ_2}{Ib_2} \quad (5.28)$$

Note: For the lower portion of the cross-section below the N.A., the Q is evaluated in the similar manner.

The typical shear stress distributions for the section plane-cuts for an "I" beam and a "T" beam are illustrated by Figures 5.11 and 5.12, respectively.

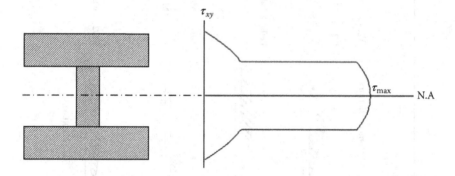

FIGURE 5.11   The "I" beam shear stress distribution.

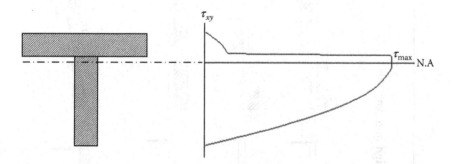

FIGURE 5.12   The "T" beam shear stress distribution.

5.7   GUIDE TO BEAM DESIGN AND ANALYSIS

TABLE 5.1
Guide to Beam Design and Analysis

F.B.D.	Shear Diagram	Moment Diagram	V_{max}	M_{max}	Deflection
Simply Supported (F at center, L/2, L/2; R=0; F/2)	F/2, L/2, L/2, −F/2	FL/4, L/2, L/2	**F/2**	**FL/4**	$\frac{Fx}{48EI}(3L^2 - 4x^2)$
Simply Supported (F, a, b; R=0; Fb/L, Fa/L)	Fb/L, a, b, −Fa/L	Fab/L, a, b	**Fa/L**	**Fab/L**	$\frac{Fbx}{6EIL}(L^2 - b^2 - x^2), x = 0.. a$ $\frac{Fa(L-x)}{6EIL}(2Lx - a^2 - x^2), x = a.. L$
Simply Supported (M, a, b; R=0; M/L, −M/L; M=(F)(h))	M/L, L	Ma/L, a, b, Mb/L	**M/L**	**Mb/L** For a<b **Ma/L**	$\frac{Mx}{6EIL}(6aL - x^2 - 3a^2 - 2L^2), x = 0.. a$ $\frac{M}{6EIL}(3a^2L - x^3 - 3L^2x - x(2L^2 + 3a^2)), a.. L$
Simply Supported (w; R=0; wL/2, wL/2)	wL/2, L/2, L/2, −wL/2	L/2, M=(wx/2)(L−x)	**wL/2**	**wL²/8**	$\frac{wx}{24EI}(L^3 - 2Lx^2 + x^3)$

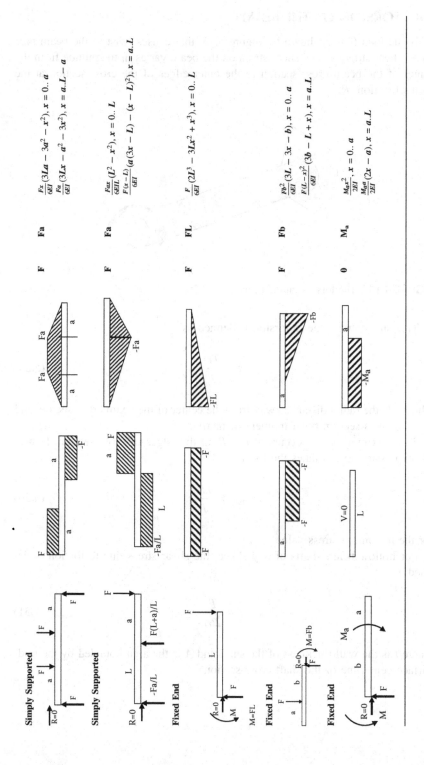

5.8 TORSION OF THE BEAMS

A torque load (T), as shown by Figure 5.13, that causes twist in the beam produces shear stresses. The shear stress on the beam varies in magnitude from the center of the beam cross-section to the outer edges of the cross-section at the radius location, R.

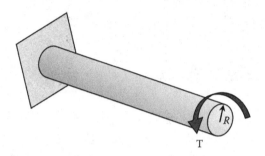

FIGURE 5.13   The torsion on the beam.

This shear stress due to torsion is defined as

$$\tau = \frac{Tr}{J} \qquad (5.29)$$

where r is the radial distance away from the center of the beam cross-section and J is the cross-section polar moment of inertia.

The maximum stress occurs at $r = R$, at the edges of cross-section. Hence, the expression (5.29) takes the form

$$\tau_{max} = \frac{TR}{J} \qquad (5.30)$$

for the maximum stress value.

For hollow beam shafts of any shape, the shear stress due to torsion is defined as

$$\tau = \frac{T}{2tA} \qquad (5.31)$$

where t is the wall thickness of the shaft and A is the area bounded by the mid-surface centerline of the shaft cross-section.

5.9   CURVED BEAM THEORY

The Winkler method is herby utilized to illustrate the curved beam theory. Consider a curved beam shown in Figure 5.14, following under an applied bending moment, M. Assume that the cross-sectional plane of the beam remains plane after deformation due to this bending. Further, let it be known that there exists a neutral axis surface whereby its deflection due to bending is zero. In Figure 5.14, this surface is a distance, R, away from the center of the beam curvature, C.

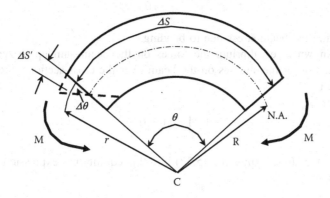

FIGURE 5.14   The curved beam under bending moment.

Now, the deformation of the curved beam, $\Delta S'$, at any location, r from the center of the curvature, can be expressed in terms of the rotational angle change, $\Delta\theta$, and the distance away from the neutral axis $(r - R)$. This deformation concept is expressed as follows:

$$\Delta S' = (r - R)\Delta\theta \tag{5.32}$$

Knowing that the original un-deflected curve of the beam at location r is

$$\Delta S = r\theta \tag{5.33}$$

one can develop an expression for the beam strain in the angular direction, θ, as

$$\varepsilon_\theta = \frac{\Delta S'}{\Delta S} \tag{5.34}$$

or, by substituting the expressions (5.32) and((5.33) into (5.34),

$$\varepsilon_\theta = \frac{(r - R)\Delta\theta}{r\theta} \tag{5.35}$$

Hence, by Hooke's principal the stress of the beam can be represented as

$$\sigma = E\varepsilon \tag{5.36}$$

whereby substituting the strain expression (5.35) into this stress equation one would have the expression

$$\sigma = E\frac{(r - R)\Delta\theta}{r\theta} \tag{5.37}$$

representing the beam stress due to bending.

It is known at equilibrium the forces on the beam sum up to zero; thus, the summation of the stresses on the beam over the cross-sectional area of the beam (A) is

$$\int \sigma dA = 0 \tag{5.38}$$

Substitute the stress expression (5.37) into this equilibrium expression and one would have

$$\int E\frac{(r - R)\Delta\theta}{r\theta}dA = 0 \tag{5.39}$$

which trivially is an indication that

$$\int \frac{(r - R)}{r}dA = 0 \tag{5.40}$$

or by separation of terms,

$$\int dA - \int \frac{R}{r}dA = 0 \tag{5.41}$$

Hence, the location of the neural axis from the curvature center, C, is determined as

$$A - R\int \frac{1}{r}dA = 0 \Rightarrow R = \frac{A}{\int \frac{dA}{r}} \tag{5.42}$$

Likewise, the moment on the beam can also be represented in terms of the stress magnitude away from the neutral axis $(r - R)$ as

$$\int (r - R)\sigma dA = M \qquad (5.43)$$

Substitute the stress expression (5.37) into this equilibrium expression, and one would have

$$\int E\frac{(r - R)^2\Delta\theta}{r\theta}dA = M \qquad (5.44)$$

which can be expanded as

$$\int E\frac{(r^2 - 2rR + R^2)\Delta\theta}{r\theta}dA = M \qquad (5.45)$$

and simplified to

$$E\frac{\Delta\theta}{\theta} \int \left(r - 2R + \frac{R^2}{r}\right)dA = M \qquad (5.46)$$

This expression can be further separated and solved for to produce

$$E\frac{\Delta\theta}{\theta}\left[\int rdA - 2R\int dA + R^2\int \frac{dA}{r}\right] = M \qquad (5.47)$$

Hence, from the elementary section properties, it is known that the centroid of a cross-section is

$$\bar{r} = \frac{\int rdA}{A} \qquad (5.48)$$

That is, by substitution of this centroid term and the expression (5.42) into expression (5.47), one would develop the expression

$$E\frac{\Delta\theta}{\theta}(\bar{r}A - 2RA + RA) = M \qquad (5.49)$$

or

$$E\frac{\Delta\theta}{\theta}(\bar{r}A - RA) = M \qquad (5.50)$$

Now, this expression is rearranged as

$$E\frac{\Delta\theta}{\theta} = \frac{M}{(\bar{r} - R)A} \tag{5.51}$$

which can be substituted back into the stress expression (5.37) to give

$$\sigma = \frac{M(r - R)}{Ar(\bar{r} - R)} \tag{5.52}$$

for the stress representation of the curved beam under the bending moment, M.

The neutral axis location for three typical cross-sections were developed here for ease calculations (Figure 5.15).

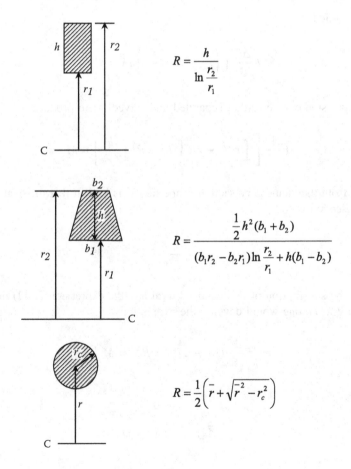

$$R = \frac{h}{\ln\dfrac{r_2}{r_1}}$$

$$R = \frac{\dfrac{1}{2}h^2(b_1 + b_2)}{(b_1 r_2 - b_2 r_1)\ln\dfrac{r_2}{r_1} + h(b_1 - b_2)}$$

$$R = \frac{1}{2}\left(\bar{r} + \sqrt{\bar{r}^2 - r_c^2}\right)$$

FIGURE 5.15 The neutral axis location for typical cross-sections.

Example 5.7: For the curved beam section below, under the 100 lbs of pulling load determine the stress at the top and bottom of the cross-section of the beam at A-A.

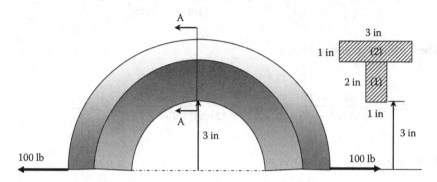

The free-body diagram of the beam section is drawn as follows.

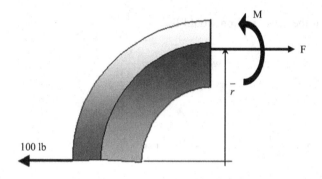

$$\sum F = 0 \Rightarrow F = 100 \ \text{lbf}$$

The centroid of the cross-section is determined first.

Section	A	$\bar{r}$	$\bar{r}A$
(1)	(1)(2) = 2 in²	4 in	8
(2)	(1)(3) = 3 in²	5.5 in	16.5
	5 in²	24.5	$\bar{r} = \frac{\sum \bar{r}A}{\sum A} = \frac{24.5}{5} = 4.9$ in

Thus, $M = -100\bar{r} = (-100)(4.9) = -490$ in-lb

The distance to the neutral axis is, $R = \frac{A}{\int \frac{dA}{r}}$; thus,

$$R = \frac{5}{\int_3^5 \frac{(1)dr}{r} + \int_5^6 \frac{(3)dr}{r}} \Rightarrow R = \frac{5}{(1)\ln\frac{5}{3} + (3)\ln\frac{6}{5}} = 4.73$$

Then, using expression (5.52),
Top: $r = 6$ in

$$\sigma_{bending} = \frac{M(r-R)}{Ar(\bar{r}-R)} \Rightarrow \sigma_{bending} = \frac{-490(6-4.73)}{(5)(6)(4.9-4.73)} = -120.2 \text{ psi}$$

Bottom: $r = 3$ in

$$\sigma_{bending} = \frac{M(r-R)}{Ar(\bar{r}-R)} \Rightarrow \sigma_{bending} = \frac{-490(3-4.73)}{(5)(3)(4.9-4.73)} = 326.8 \text{ psi}$$

Now, the total stress on the cross-section would be:

$$\sigma = \sigma_{axial} + \sigma_{bending} \quad \text{where,} \quad \sigma_{axial} = \frac{F}{A}$$

At the top:

$$\sigma = \frac{100}{5} - 120.2 = -100.2 \text{ psi}$$

At the bottom:

$$\sigma = \frac{100}{5} + 325.8 = 345.8 \text{ psi}$$

Example 5.7: (S.I. Units)

For the curved beam section below, under the 444.82 N of pulling load determine the stress at the top and bottom of the cross-section of the beam at A-A.

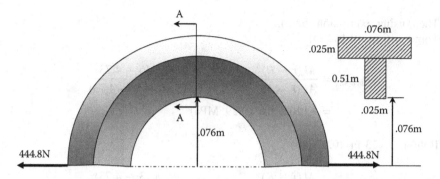

The free-body diagram of the beam section is drawn as follows.

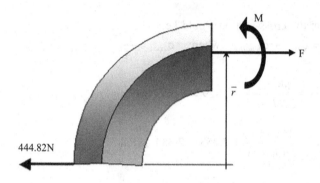

$$\sum F = 0 \Rightarrow F = 444.82N$$

The centroid of the cross-section is determined first.

Section	A	$\bar{r}$	$\bar{r}A$
(1)	(1)(2) = 2 in.^{2}	4 in.	8
(2)	(1)(3) = 3 in.^{2}	5.5 in.	16.5
	5 in.^{2}	(3.226E – 3 m^2)	24.5

$$\bar{r} = \frac{\sum \bar{r}A}{\sum A} = \frac{24.5}{5} = 4.9 \text{ in.}$$
$$(0.124 \text{ m})$$

Thus, $M = -100\bar{r} = (-444.8)(.124) = -55.2N - m$

The distance to the neutral axis is, $R = \frac{A}{\int \frac{dA}{r}}$; thus,

$$R = \frac{5}{\int_3^5 \frac{(1)dr}{r} + \int_5^6 \frac{(3)dr}{r}} \Rightarrow R = \frac{5}{(1)\ln \frac{5}{3} + (3)\ln \frac{6}{5}} = 4.73$$

Then, using expression (5.52),

Top: $r = 6$ in. (0.152 m)

$$\sigma_{bending} = \frac{M(r - R)}{Ar(\bar{r} - R)} \Rightarrow \sigma_{bending} = \frac{-490(6 - 4.73)}{(5)(6)(4.9 - 4.73)}$$

$$= -120.2 \text{ psi } (.829 \text{ MPa})$$

Bottom: $r = 3$ in. (0.076 m)

$$\sigma_{bending} = \frac{M(r - R)}{Ar(\bar{r} - R)} \Rightarrow \sigma_{bending} = \frac{-490(3 - 4.73)}{(5)(3)(4.9 - 4.73)}$$

$$= 326.8 \text{psi } (2.253 \text{ MPa})$$

Now, the total stress on the cross-section would be:

$$\sigma = \sigma_{axial} + \sigma_{bending} \quad \text{where,} \quad \sigma_{axial} = \frac{F}{A}$$

At the top:

$$\sigma = \frac{444.82}{3.226E - 3} - .829 = -.691 \text{ MPa}$$

At the bottom:

$$\sigma = \frac{444.82}{3.226E - 3} + 2.253 = 2.384 \text{ MPai}$$

Problems

1. For the beam system shown, determine the shear and moment diagram.

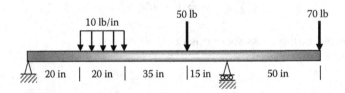

2. For the beam system shown, determine the shear and moment diagram.

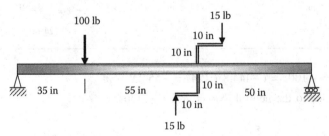

3. Determine the deflection of this beam under the applied loading. (Assume E and I properties.)

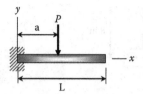

4. Determine the deflection of this beam under the applied loading. (Assume E and I properties.)

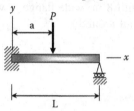

5. Determine the deflection of this beam under the applied loading. (Assume E and I properties.)

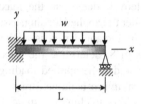

6. Determine the deflection of this beam under the applied loading. (Assume E and I properties.)

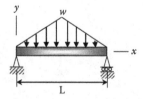

7. Determine the deflection of this beam under the applied loading. (Assume E and I properties.)

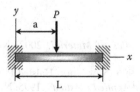

8. Determine the maximum deflection of this steel $(E = 29 \times 10^6$ psi) beam under the applied loading.

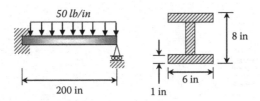

9. For the beam system shown, determine the maximum bending and shear stress of this steel ($E = 29 \times 10^6$ psi) beam under the applied loading. (Assume a T cross-section with 8 in. wide flange, a web of 10 in. high and a thickness of 1.5 in. all around.)

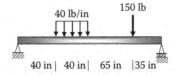

10. For the beam system shown in problem 1, determine the maximum bending and shear stress of this beam under the applied loading, assuming an aluminum material and a 2 in. by 3 in. rectangular cross-section.
11. For the beam system shown in problem 8, determine the maximum bending and shear stress of this beam under the applied loading.
12. For the hook shown, determine the maximum bending stress and its corresponding location under the applied loading (Assume E and I properties.) Also, size the cross-section of the hook if an aluminum material is to be used.

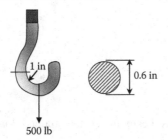

REFERENCES

Beer, F.P., Johnston, E.R., DeWolf, J.T., *Mechanics of Material*, 2002. New York: McGraw Hill.
Megson, T.H.G., *Structural and Stress Analysis*, 1996. New York: Halsted Press.
Ugural, A.C., Fenster, S.K., *Advanced Strength and Applied Elasticity*, 1995. New Jersey: Prentice Hall.

6 Plate Analysis Theory

6.1 INTRODUCTION

This chapter covers the analysis of initially flat circular and rectangular plates. To be specific, thin plates are analyzed where the ratio of the thickness to the smaller length is less than 1/20th. The deflection behavior of the plate is shown, and in addition the in-plane stress formulations are developed. Timoshenko's approach is taken in this chapter. The general assumptions used for plate analysis are: 1) the plate deflection of the mid-surface is small compared to the plate thickness, 2) the lines normal to the mid-surface remain straight and normal to the mid-surface after the bending of the plates and 3) the thickness stress is negligible.

6.2 CIRCULAR PLATES UNIFORMLY LOADED

For a circular plate of radius "a" shown by Figure 6.1, loaded by a uniformly distributed load on the surface, the equation of the deflection is derived by Timoshenko (1959) as

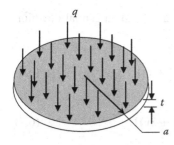

FIGURE 6.1 The uniformly loaded circular plate.

$$w = \frac{qr^4}{64D} + C1\frac{r^2}{4} + C2 \log \frac{r}{a} + C3 \tag{6.1}$$

where q is the pressure distribution (lb/in^2), r is the radius location and D is the stiffness of the plate defined as $D = \frac{Et^3}{12(1 - v^2)}$.

The constants of this equation can be solved for by determining the deflection and the slope of the circular plate for different edge boundary conditions.

Consider a uniformly loaded plate clamped at the edges, as shown by Figure 6.2.

DOI: 10.1201/9781003311218-6

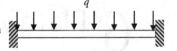

FIGURE 6.2 Uniformly loaded circular plate with clamped edges.

The following boundary conditions apply for this plate:

$$\text{At } r = 0 \text{ and } r = a, \quad \frac{dw}{dr} = 0 \tag{6.2}$$

$$\text{At } r = a, \quad w = 0 \tag{6.3}$$

Using the above boundary conditions, expression (6.1) can be manipulated to determine the equation constants, C1, C2 and C3. Thus, resulting in the general form for the deflection behavior of a clamped-edge plate.

$$w = \frac{q}{64D}(a^2 - r^2)^2 \tag{6.4}$$

It is known that the maximum deflection occurs at the center of the plate. Hence, at $r = 0$ the maximum plate deflection is

$$w_{\text{max}} = \frac{qa^4}{64D} \tag{6.5}$$

Further, the bending moment for the plate in the radial and tangential direction are given as

$$M_r = \frac{q}{16}[a^2(1 + v) - r^2(3 + v)] \tag{6.6}$$

$$M_t = \frac{q}{16}[a^2(1 + v) - r^2(1 + 3v)] \tag{6.7}$$

Based on the bending moments, the radial and tangential stresses in the plate are given by

$$\sigma_r = \frac{6M_r}{t^2} \text{ or } \sigma_r = \frac{6q}{16t^2}[a^2(1 + v) - r^2(3 + v)] \tag{6.8}$$

$$\sigma_t = \frac{6M_t}{t^2} \text{ or } \sigma_t = \frac{6q}{16t^2}[a^2(1 + v) - r^2(1 + 3v)] \tag{6.9}$$

The radial and tangential stresses at the edges of the plate where $r = a$ are

$$\sigma_{r,r=a} = \frac{6q}{16t^2}[a^2(1 + v) - a^2(3 + v)] = \frac{3a^2q}{4t^2} \tag{6.10}$$

where this is also the maximum radial stress on the plate and

$$\sigma_{t,r=a} = \frac{6q}{16t^2}[a^2(1 + v) - a^2(1 + 3v)] = \frac{3a^2vq}{4t^2} \tag{6.11}$$

where the maximum tangential stress would be at $r = 0$.

Now, consider a uniformly loaded plate simply supported at the edges, as shown by Figure 6.3, below.

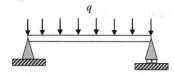

FIGURE 6.3 Uniformly loaded circular plate with simply supported edges.

The following boundary conditions apply for this plate:

$$\text{At } r = 0 \quad \frac{dw}{dr} = 0 \tag{6.12}$$

$$\text{At } r = a, \quad w = 0 \tag{6.13}$$

Applying the above boundary conditions to expression (6.1), the deflection behavior of the simply supported circular plate is determined as

$$w = \frac{q}{64D}(a^2 - r^2)\left(\frac{5 + v}{1 + v}a^2 - r^2\right) \tag{6.14}$$

Once again the maximum deflection occurs at the center of the plate at $r = 0$. Thus, expression (6.14) reduces to the following form for the maximum plate deflection:

$$w_{max} = \frac{qa^4}{64D}\left(\frac{5 + v}{1 + v}\right) \tag{6.15}$$

The bending moments in the radial and tangential directions for this plate are given as

$$M_r = \frac{q}{16}(a^2 - r^2)(3 + v) \tag{6.16}$$

$$M_t = \frac{q}{16}[a^2(3 + v) - r^2(1 + 3v)] \qquad (6.17)$$

which produce the following form for the stresses on the plate in the radial and tangential directions, respectively.

$$\sigma_r = \frac{6M_r}{t^2} \text{ or } \sigma_r = \frac{6q}{16t^2}(a^2 - r^2)(3 + v) \qquad (6.18)$$

$$\sigma_t = \frac{6M_t}{t^2} \text{ or } \sigma_t = \frac{6q}{16t^2}[a^2(3 + v) - r^2(1 + 3v)] \qquad (6.19)$$

The maximum stresses are known to occur at the center of the plate at $r = 0$ for this type of plate, as shown by expressions (6.20) and (6.21) below:

$$\sigma_{r,\max} = \frac{6a^2q}{16t^2}(3 + v) \qquad (6.20)$$

$$\sigma_{t,\max} = \frac{6a^2q}{16t^2}(3 + v) \qquad (6.21)$$

6.3   CIRCULAR PLATES LOADED AT THE CENTER

Likewise, in a similar manner a circular plate with concentrated load at the center is analyzed. For a circular plate of radius "a" shown by Figure 6.4, loaded by a concentrated load at the center, the equation of the deflection is derived as

$$w = \frac{P}{8\pi D}r^2 \log \frac{r}{a} + C1r^2 + C2 \qquad (6.22)$$

where P is the load at the center (lb), r is the radius location and D is the stiffness of the plate defined as $D = \frac{Et^3}{12(1 - v^2)}$.

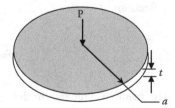

FIGURE 6.4   The center-loaded circular plate.

The constants of this equation can also be solved for by determining the deflection and the slope of the circular plate for different edge boundary conditions.

Now, consider a circular center-loaded plate, clamped at the edges as shown by Figure 6.5 above.

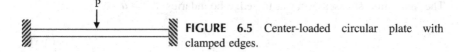

FIGURE 6.5 Center-loaded circular plate with clamped edges.

The following boundary conditions apply for this plate:

$$\text{At } r = 0 \quad \frac{dw}{dr} = 0 \tag{6.23}$$

$$\text{At } r = a, \quad w = 0 \tag{6.24}$$

Applying the above boundary conditions to expression (6.22), the deflection behavior of this simply supported circular plate is determined as

$$w = \frac{Pr^2}{8\pi D} \log \frac{r}{a} + \frac{P}{16\pi D}(a^2 - r^2) \tag{6.25}$$

Likewise, as the plate described in section 6.2, the maximum deflection occurs at the center of the plate at $r = 0$. Hence, expression (6.25) reduces to the following form for the maximum plate deflection at $r = 0$

$$w_{\max} = \frac{Pa^2}{16\pi D} \tag{6.26}$$

The radial and tangential bending moments for this type of plate are defined as

$$M_r = \frac{P}{4\pi}\left[(1 + v)\log \frac{a}{r} - 1\right] \tag{6.27}$$

$$M_t = \frac{P}{4\pi}\left[(1 + v)\log \frac{a}{r} - v\right] \tag{6.28}$$

Thus, again the stresses in the radial and tangential direction are derived as follows, respectively,

$$\sigma_r = \frac{6M_r}{t^2} \text{ or } \sigma_r = \frac{6P}{4\pi t^2}\left[(1 + v)\log \frac{a}{r} - 1\right] \tag{6.29}$$

$$\sigma_t = \frac{6M_t}{t^2} \text{ or } \sigma_t = \frac{6P}{4\pi t^2}\left[(1 + v)\log\frac{a}{r} - v\right] \tag{6.30}$$

The maximum stresses occur at the edge boundaries at $r = a$

$$\sigma_{r,\max} = -\frac{6P}{4\pi t^2} \tag{6.31}$$

$$\sigma_{t,\max} = -\frac{6Pv}{4\pi t^2} \tag{6.32}$$

For a center-loaded plate, simply supported at the edges, as shown by Figure 6.6,

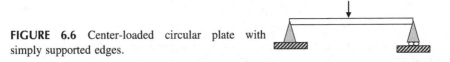

FIGURE 6.6 Center-loaded circular plate with simply supported edges.

the equation of the deflection is derived as

$$w = \frac{P}{16\pi D}\left[\frac{3 + v}{1 + v}(a^2 - r^2) + 2r^2 \log\frac{r}{a}\right] \tag{6.33}$$

Once again the maximum deflection occurs at the plate center at $r = 0$ for this type of loading as well. Expression (6.33) reduces to the following form, representing the maximum plate deflection,

$$w_{\max} = \frac{Pa^2}{16\pi D}\frac{(3 + v)}{(1 + v)} \tag{6.34}$$

The radial and tangential bending moments for this type of plate are defined as

$$M_r = \frac{P}{4\pi}(1 + v)\log\frac{a}{r} \tag{6.35}$$

$$M_t = \frac{P}{4\pi}\left[(1 + v)\log\frac{a}{r} + 1 - v\right] \tag{6.36}$$

which, respectively, produce the following stresses in the radial and tangential directions for a plate simply supported and point loaded at the center:

$$\sigma_r = \frac{6M_r}{t^2} \text{ or } \sigma_r = \frac{6P}{4\pi t^2}(1 + v)\log\frac{a}{r} \tag{6.37}$$

$$\sigma_t = \frac{6M_t}{t^2} \text{ or } \sigma_t = \frac{6P}{4\pi t^2}\left[(1 + v)\log\frac{a}{r} + 1 - v\right] \qquad (6.38)$$

Example 6.1: A flat steel circular plate of 0.06 in. thick with diameter of 6 in. is subjected to a 6 psi uniform pressure on the surface. Determine the maximum plate deflection at the center and the radial stresses at the center and boundary edges of the plate ($E = 29 \times 10^6$ psi, $v = 0.3$).

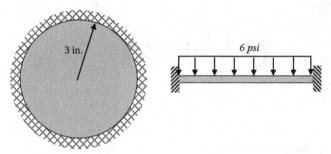

3 in.

6 psi

The stiffness of the plate is determined as

$$D = \frac{Et^3}{12(1 - v^2)}, \quad D = \frac{(29000000)(.06)^3}{12(1 - .3^2)} = 573.6$$

The maximum deflection of the plate at the center is determined by expression (6.5) as follows:

$$w_{max} = \frac{qa^4}{64D}, \quad w_{max} = \frac{(6)(3)^4}{64(573.6)} = 0.013 \text{ in.}$$

The radial stresses are determined by using expression (6.8).

$$\sigma_r = \frac{6q}{16t^2}[a^2(1 + v) - r^2(3 + v)]$$

The stress at the center of the plate at $r = 0$ is

$$\sigma_{r=0} = \frac{6(6)}{16(.06)^2}[(3)^2(1 + .3) - (0)^2(3 + .3)] = 7312.5 \text{ psi}$$

The stress at the boundary edges of the plate at $r = 3$ in. is

$$\sigma_{r=3} = \frac{6(6)}{16(.06)^2}[(3)^2(1 + .3) - (3)^2(3 + .3)] = -11250 \text{ psi}$$

Example 6.1: (S.I. Units)

A flat steel circular plate of 1.524E-3 m thick with diameter of 0.152 m is subjected to a 0.041 MPa uniform pressure on the surface. Determine the maximum plate deflection at the center and the radial stresses at the center and boundary edges of the plate. ($E = 2.0 \times 10^5 MPa$, $v = 0.3$)

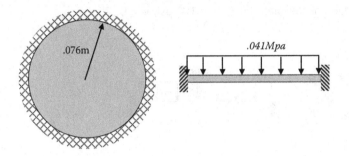

The stiffness of the plate is determined as

$$D = \frac{Et^3}{12(1 - v^2)}, \quad D = \frac{(2.0E11)(1.524E - 3)^3}{12(1 - v^2)}$$

$$= 64.83 \text{ (use consistent units)}$$

The maximum deflection of the plate at the center is determined by expression (6.5), as shown below:

$$w_{max} = \frac{qa^4}{64D}$$

$$wmax = \frac{(.041E6)(.076)^4}{64(64.83)} = 3.297E - 4m$$

$$wmax = \frac{(.041E6)(.076)^4}{64(6.483E - 5)} = \text{(use consistent units)}$$

The radial stresses are determined by using the expression (6.8).

$$\sigma_r = \frac{6q}{16t^2}[a^2(1 + v) - r^2(3 + v)]$$

The stress at the center of the plate at $r = 0$ is

$$\sigma_{r=0} = \frac{6(.041E6)}{16(1.524E - 3)^2}[(.076)^2(1 + .3) - (0)^2(3 + .3)] = 50.4 \text{ MPa}$$

The stress at the boundary edges of the plate at $r = .076m$ is

$$\sigma_{r=3} = \frac{6(.041E6)}{16(1.524E - 3)^2}[(.076)^2(1 + .3) - (.076)^2(3 + .3)] = 77.6\text{MPa}$$

6.4   UNIFORMLY LOADED RECTANGULAR PLATES

Similarly, the behavior of rectangular plates can be shown by plate theory of elasticity. Here, the essentials of this method are described for rectangular plates. Consider a rectangular plate simply supported at all edges with sides "a" and "b" as shown by Figure 6.7, below,

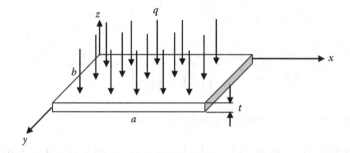

FIGURE 6.7   Rectangular plate under uniform loading.

The plate is uniformly loaded on the surface. Fourier series is used to represent the deflection of this plate as

$$w = \frac{16q}{\pi^6 D} \sum_{m=1}^{\infty} \sum_{n=1}^{\infty} \frac{\sin \frac{m\pi x}{a} \sin \frac{n\pi y}{b}}{mn\left(\frac{m^2}{a^2} + \frac{n^2}{b^2}\right)^2} \qquad (6.39)$$

where m = 1,3,5 … .. and n = 1,3,5… …
    The maximum deflection for this rectangular plate occurs at $x = a/2$ and $y = b/2$. Thus, by applying these coordinates, the expression (6.39) is reduced to the following form to show the maximum plate deflection as

$$w_{\max} = \frac{16q}{\pi^6 D} \sum_{m=1}^{\infty} \sum_{n=1}^{\infty} \frac{(-1)^{\frac{m+n}{2}-1}}{mn\left(\frac{m^2}{a^2} + \frac{n^2}{b^2}\right)^2} \qquad (6.40)$$

where m = 1,3,5 … .. and n = 1,3,5… …

The stresses in the "x" and "y" directions are evaluated as

$$\sigma_x = \frac{6D}{h^2}\left(\frac{\partial^2 w}{\partial x^2} + v\frac{\partial^2 w}{\partial y^2}\right) \tag{6.41}$$

$$\sigma_y = \frac{6D}{h^2}\left(\frac{\partial^2 w}{\partial y^2} + v\frac{\partial^2 w}{\partial x^2}\right) \tag{6.42}$$

where application of the deflection relation (6.40) into expressions (6.41) and (6.42) above, the stress field in the rectangular plate is defined as

$$\sigma_x = \frac{96q}{h^2\pi^4}\sum_{m=1}^{\infty}\sum_{n=1}^{\infty}\frac{\left(\frac{m}{a}\right)^2 + v\left(\frac{n}{b}\right)^2}{mn\left[\left(\frac{m}{a}\right)^2 + \left(\frac{n}{b}\right)^2\right]^2}\sin\frac{m\pi x}{a}\sin\frac{m\pi y}{b} \tag{6.43}$$

$$\sigma_y = \frac{96q}{h^2\pi^4}\sum_{m=1}^{\infty}\sum_{n=1}^{\infty}\frac{v\left(\frac{m}{a}\right)^2 + \left(\frac{n}{b}\right)^2}{mn\left[\left(\frac{m}{a}\right)^2 + \left(\frac{n}{b}\right)^2\right]^2}\sin\frac{m\pi x}{a}\sin\frac{m\pi y}{b} \tag{6.44}$$

in the "x" and "y" directions, respectively.

Example 6.2: For a steel rectangular plate of 10 in. wide by 20 in. long, with thickness of 0.1, plot the deflection of the plate along the length of the plate at the width of 5 in. There is a 1 psi uniform load on the plate surface. Assume the plate is simply supported in all edges.

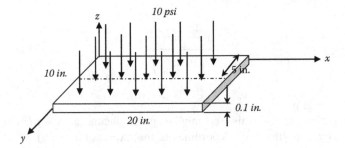

The stiffness of the plate is determined as:

$$D = \frac{Et^3}{12(1-v^2)}, \quad D = \frac{(29000000)(0.1)^3}{12(1-.3^2)} = 2655.7$$

Using the expression 6.42, the deflection in terms of the x at $y = 5$ in. is, (using the first terms for the series, $m = n = 1$):

$$w = \frac{16q}{\pi^6 D} \sum_{m=1}^{\infty} \sum_{n=1}^{\infty} \frac{\sin\frac{m\pi x}{a}\sin\frac{n\pi y}{b}}{mn\left(\frac{m^2}{a^2} + \frac{n^2}{b^2}\right)^2}, \quad w(x, 5)$$

$$= \frac{16(1)}{\pi^6 2655.7} \left[\frac{\sin\frac{(1)\pi x}{20}\sin\frac{(1)\pi(5)}{10}}{(1)(1)\left(\frac{1^2}{20^2} + \frac{1^2}{10^2}\right)^2} \right]$$

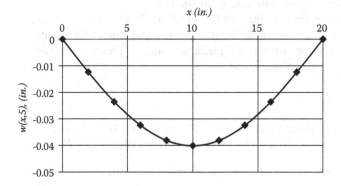

Problems

1. A flat steel circular plate of 0.06 in. thick with a diameter of 6 in. is subjected to a 6 psi uniform pressure on the surface. Assuming the edges of the plate are simply supported, determine the maximum plate deflection at the center.

2. A flat aluminum 6061-T6 circular plate of 0.1 in. thick with diameter of 10 in. is subjected to a 15 psi uniform pressure on the surface. Assuming the edges of the plate are fixed, determine the maximum plate deflection at the center and the maximum stress on the plate.

3. Consider a flat circular plate of 0.05 in. thick with a diameter of 12 in. made out of aluminum 6061-T6. Assuming the edges of the plate are fixed, what is the maximum surface pressure it could withstand before yielding occurs?

4. Consider the plate in problem 3. Assuming the edges of the plate are simply supported, what is the maximum surface pressure it could withstand before yielding occurs?

5. Consider a flat circular plate of 0.04 in. thick with a diameter of 8 in. made out of aluminum 2014-T6. The plate is subjected to a concentrated load of 100 lb at the center. Assuming the edges of the plate are fixed, determine the maximum plate deflection at the center and the maximum stress on the plate.

6. Do problem 5 above but assume the edges are simply supported.

7. Consider a flat circular plate of 0.15 in. thick with a diameter of 15 in. made out of stainless steel. Assuming the edges of the plate are fixed, what is the maximum surface pressure it could withstand before the plate fails?

8. Consider a flat square plate of 0.15 in. thick with sides of 25 in. long made out of stainless steel. The plate is subjected to a constant surface pressure of 20 psi. Assuming the edges of the plate are simply supported, what would be the maximum plate deflection? Determine the maximum stress on the plate under this loading.

9. Consider a flat rectangular plate of 0.1 in. thick, 25 in. wide by 40 in. long made out of aluminum 7075-T6. Assuming the edges of the plate are simply supported, what is the maximum surface pressure it could withstand before yielding occurs?

REFERENCES

Megson, T.H.G., *Structural and Stress Analysis*, 1996. New York: Halsted Press.

Timoshenko, S.P., Woinowsky-Krieger, S., *Theory of Plates and Shells*, 1959. New York: McGraw Hill.

Ugural, A.C., Fenster, S.K., *Advanced Strength and Applied Elasticity*, 1995. New Jersey: Prentice Hall.

7 Elastic Stability and Buckling

7.1 INTRODUCTION

Typically, a structural member is said to have buckled when the member fails to react to the bending moment generated by a compressive load on it. In this chapter, the buckling of long slender columns and thin plates will be considered. A "long" column may be assumed to be one having a slenderness ratio, or length to radius of gyration ratio, greater than 200. Radius of gyration is defined as the square root of the column second moment of area over the cross-sectional area of the column. Likewise, a plate is called "thin" plate when its thickness is at least one order of magnitude smaller than the span or diameter of the plate.

7.2 COLUMN BUCKLING INSTABILITY

The following mathematical formulation, as used by Simitses (1976), represents the behavior of a column of length L, under an axial loading, P, parallel to its axis without eccentricity,

$$EI\frac{d^4u}{dx^4} + P\frac{d^2u}{dx^2} = 0 \tag{7.1}$$

where "u" is the lateral deflection of the column perpendicular to the column axis.

Rewriting this equation in a simpler form, one would have:

$$\frac{d^4u}{dx^4} + \frac{P}{EI}\frac{d^2u}{dx^2} = 0 \tag{7.2}$$

Introducing the term,

$$k^2 = \frac{P}{EI} \tag{7.3}$$

equation (7.2) above reduces to

$$\frac{d^4u}{dx^4} + k^2\frac{d^2u}{dx^2} = 0 \tag{7.4}$$

The available solution for this differential equation is

$$u = u(x) = c_1 \sin kx + c_2 \cos kx + c_3 x + c_4 \tag{7.5}$$

which, at application of different boundary conditions and expansion of the solution, provides the critical loading of the column buckling ("x" is in the direction along the length).

Now, consider a long slender column as shown by Figure 7.1, under axial loading, P. The following boundary conditions are applicable to this system,

$$
\begin{aligned}
u(A) &= u(0) = 0 \\
u(B) &= u(L) = 0 \\
\frac{d^2u}{dx^2}(A) &= \frac{d^2u}{dx^2}(0) = 0 \\
\frac{d^2u}{dx^2}(B) &= \frac{d^2u}{dx^2}(L) = 0
\end{aligned}
\tag{7.6}
$$

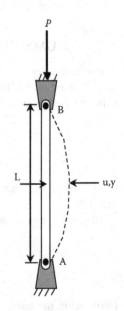

FIGURE 7.1   Simply-simply supported column.

Substituting the boundary conditions into the general solution (7.5), one would have

$$c_2 + c_4 = 0 \tag{7.7}$$

$$c_1 \sin kL + c_2 \cos kL + c_3 L + c_4 = 0 \tag{7.8}$$

$$-c_2 k^2 = 0 \tag{7.9}$$

$$-c_1 k^2 \sin kL - c_2 k^2 \cos kL = 0 \tag{7.10}$$

By rewriting the above equations in matrix form, one would have:

$$\begin{bmatrix} 0 & 1 & 0 & 1 \\ \sin kL & \cos kL & L & 1 \\ 0 & -k^2 & 0 & 0 \\ -k^2 \sin kL & -k^2 \cos kL & 0 & 0 \end{bmatrix} \begin{Bmatrix} c_1 \\ c_2 \\ c_3 \\ c_4 \end{Bmatrix} = \begin{Bmatrix} 0 \\ 0 \\ 0 \\ 0 \end{Bmatrix} \tag{7.11}$$

Once the determinant of this matrix is determined, then one would have:

$$\sin kL = 0 \tag{7.12}$$

which has the first-term solution of

$$kL = \pi \tag{7.13}$$

Substitute back the "k" loading term from expression (7.3) into equation (7.13) and one would develop the critical loading where the buckling occurs as

$$\sqrt{\frac{P_{cr}}{EI}} L = \pi \ \text{ or } \ P_{cr} = \frac{\pi^2 EI}{L^2} \tag{7.14}$$

The corresponding stress would be

$$\sigma_{cr} = \frac{P_{cr}}{A} \ \text{ or } \ \sigma_{cr} = \frac{\pi^2 EI}{AL^2} \tag{7.15}$$

Similarly, for a column fixed at both ends, as shown by Figure 7.2, the following boundary conditions are applicable:

$$\begin{aligned} u(A) &= u(0) = 0 \\ u(B) &= u(L) = 0 \\ \frac{du}{dx}(A) &= \frac{du}{dx}(0) = 0 \\ \frac{du}{dx}(B) &= \frac{du}{dx}(L) = 0 \end{aligned} \tag{7.16}$$

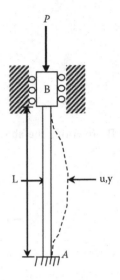

FIGURE 7.2 Fixed-fixed supported column.

By substituting the boundary conditions (7.16) into the general solution (7.5), one would have

$$
\begin{bmatrix}
0 & 1 & 0 & 1 \\
\sin kL & \cos kL & L & 1 \\
k & 0 & 1 & 0 \\
k\cos kL & -k\sin kL & 1 & 0
\end{bmatrix}
\begin{Bmatrix}
c_1 \\ c_2 \\ c_3 \\ c_4
\end{Bmatrix}
=
\begin{Bmatrix}
0 \\ 0 \\ 0 \\ 0
\end{Bmatrix}
\tag{7.17}
$$

Likewise, expanding the determinant of this matrix yields

$$
\sin \frac{kL}{2} = 0 \tag{7.18}
$$

which has the first-term solution of

$$
\frac{kL}{2} = \pi \tag{7.19}
$$

Substitute back the "k" term from expression (7.3) and one would have

$$
\sqrt{\frac{P_{cr}}{EI}}\,\frac{L}{2} = \pi \quad \text{or} \quad P_{cr} = \frac{4\pi^2 EI}{L^2} \tag{7.20}
$$

The corresponding stress would be

$$
\sigma_{cr} = \frac{P_{cr}}{A} \quad \text{or} \quad \sigma_{cr} = \frac{4\pi^2 EI}{AL^2} \tag{7.21}
$$

Finally, for a column with one end fixed and one end free, as shown by Figure 7.3, the following boundary conditions are applicable:

$$u(A) = u(0) = 0$$
$$\frac{du}{dx}(A) = \frac{du}{dx}(0) = 0$$
$$\frac{d^2u}{dx^2}(B) = \frac{d^2u}{dx^2}(L) = 0 \tag{7.22}$$
$$k^2\frac{du}{dx}(B) + \frac{d^4u}{dx^4}(B) = k^2\frac{du}{dx}(L) + \frac{d^4u}{dx^4}(L) = 0$$

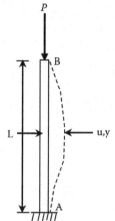

FIGURE 7.3 Fixed-free supported column.

By substituting the boundary conditions (7.22) into the general solution (7.5), one would have

$$\begin{bmatrix} 0 & 1 & 0 & 1 \\ k & 0 & 1 & 0 \\ -k^2\sin kL & -k^2\cos kL & 0 & 0 \\ 0 & 0 & k^2 & 0 \end{bmatrix}\begin{Bmatrix} c_1 \\ c_2 \\ c_3 \\ c_4 \end{Bmatrix} = \begin{Bmatrix} 0 \\ 0 \\ 0 \\ 0 \end{Bmatrix} \tag{7.23}$$

Expanding the determinant of this matrix yields

$$\cos kL = 0 \tag{7.24}$$

which has the first-term solution of

$$kL = \frac{\pi}{2} \tag{7.25}$$

Substitute back the "k" term from expression (7.3) and one would have

$$\sqrt{\frac{P_{cr}}{EI}} L = \frac{\pi}{2} \quad \text{or} \quad P_{cr} = \frac{\pi^2 EI}{4L^2} \tag{7.26}$$

The corresponding stress would be

$$\sigma_{cr} = \frac{P_{cr}}{A} \quad \text{or} \quad \sigma_{cr} = \frac{\pi^2 EI}{4AL^2} \tag{7.27}$$

Example 7.1: For the simply supported column of 10 ft. long with square cross-section of 3 in. by 3 in. determine the critical buckling stress. Assume a steel column.

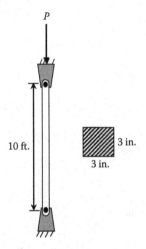

Using expression (7.15), the critical stress of the column is determined as

$$\sigma_{cr} = \frac{\pi^2 EI}{AL^2}$$

$$E = 29 \times 10^6 \ psi$$

$$A = 3 \times 3 = 9 \ in^2$$

$$I = \frac{3 \times 3^3}{12} = 6.75 \ in^4$$

$$\sigma_{cr} = \frac{\pi^2 (29 \times 10^6)(6.75)}{(9)(10 * 12)^2} = 14906 \ psi$$

Compare this value to the steel yield strength.

7.3 COLUMN BUCKLING UNDER COMBINED AXIAL AND BENDING LOADS

Consider the column loaded with a combination of axial and bending loads, as shown by Figure 7.4.

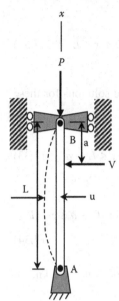

FIGURE 7.4 The column buckling under combined loading.

The equation representing the bending of the column is

$$EI\frac{d^2u}{dx^2} = -M \tag{7.28}$$

By virtue of simple static theory, the lateral reactions at points A and B are $R_A = \frac{Va}{L}$ and $R_B = \frac{V(L-a)}{L}$, respectively. Using these reaction loads, one can develop the total bending moment along the axis of the column as

$$M_1 = Pu_1 + \frac{Vax}{L} \quad \text{for } 0 \le x \le L - a \tag{7.29}$$

and

$$M_2 = Pu_2 + \frac{V(L-a)(L-x)}{L} \quad \text{for } L - a \le x \le L \tag{7.30}$$

Now, by substituting the moment equations (7.29) and (7.30) into expression (7.28), one would have

$$EI\frac{d^2u_1}{dx^2} = -Pu_1 - \frac{Vax}{L} \quad \text{for } 0 \leq x \leq L - a \quad (7.31)$$

$$EI\frac{d^2u_2}{dx^2} = -Pu_2 - \frac{V(L-a)(L-x)}{L} \quad \text{for } L - a \leq x \leq L \quad (7.32)$$

As before, by introduction of the term $k^2 = \frac{P}{EI}$, the available solutions for these differential equations are

$$u_1 = u_1(x) = c_1 \cos kx + c_2 \sin kx - \frac{Vax}{PL} \quad \text{for } 0 \leq x \leq L - a \quad (7.33)$$

$$u_2 = u_2(x) = c_3 \cos kx + c_4 \sin kx - \frac{V(L-a)(L-x)}{PL} \quad \text{for } L - a \leq x \leq L \quad (7.34)$$

For this system, by applying the applicable boundary conditions that are known as before,

$$\begin{aligned}
u_1(A) &= u_1(0) = 0 \\
u_2(A) &= u_2(0) = 0 \\
u_1(L-a) &= u_2(L-a) \\
\frac{du_1(L-a)}{dx} &= \frac{du_2(L-a)}{dx}
\end{aligned} \quad (7.35)$$

the constants of the integration c_1, c_2, c_3 and c_4 are determined as follows:

$$c_1 = 0 \quad (7.36)$$

$$c_2 = \frac{V \sin ka}{Pk \sin kL} \quad (7.37)$$

$$c_3 = \frac{V \sin k(L-a)}{Pk} \quad (7.38)$$

$$c_4 = -\frac{V \sin k(L - a)}{Pk \tan kL} \tag{7.39}$$

By substitution of the above constants into expressions (7.33) and (7.34), one would have the following final deflection expressions for the column:

$$u_1 = \frac{V \sin ka}{Pk \sin kL} \sin kx - \frac{Va}{PL}x \text{ for } 0 \le x \le L - a \tag{7.40}$$

$$u_2 = \frac{V \sin k(L - a)}{Pk \sin kL} \sin k(L - x) - \frac{V(L - a)(L - x)}{PL} \text{ for } L - a \le x \le L \tag{7.41}$$

It is known that the maximum lateral displacement occurs in the middle of the column, where $x = \frac{L}{2}$. Thus, using expressions (7.40) and (7.41), for the maximum displacement one would have (for $a = \frac{L}{2}$)

$$u\left(x = \frac{L}{2}\right) = \frac{V \sin^2 k\frac{L}{2}}{Pk \sin kL} - \frac{VL}{4P} \tag{7.42}$$

We know that expression (7.42) goes to infinity at $sin\ kL = 0$. Thus, the first-term solution is

$$kL = \pi \tag{7.43}$$

which, by substitution of the k term, would produce the critical axial load for buckling of the column as

$$\sqrt{\frac{P_{cr}}{EI}} = \frac{\pi}{L} \text{ or } P_{cr} = \frac{\pi^2 EI}{L^2} \tag{7.44}$$

7.4 MULTIPLE COLUMN SYSTEM BUCKLING

At the column system shown by Figure 7.5, points A, B and D are pinned and point C is a fixed joint. Knowing that the deflection angles produced at point C are equal and the summation of the moments at that point result to zero, one can derive the critical buckling load for the system as

$$P_{cr} = \frac{13.84EI}{L^2} \tag{7.45}$$

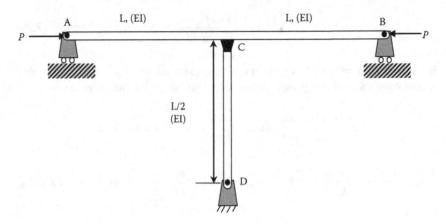

FIGURE 7.5   The three-column-system with pinned support.

The final buckling deflection of this system is shown in Figure 7.6 below.

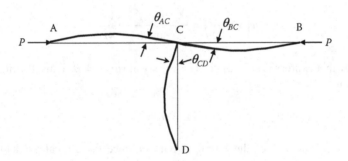

FIGURE 7.6   The three-column-system buckling deformation.

Likewise, in the column system shown in Figure 7.7, points A and B are pinned and point D is fixed. Also, point C is a fixed joint. Knowing that the deflection angles produced at point C are equal, the summation of the moments at that point result to zero, and that the deflection angle at point D is zero, one can derive the critical buckling load for the system as

$$P_{cr} = \frac{13EI}{L^2} \tag{7.46}$$

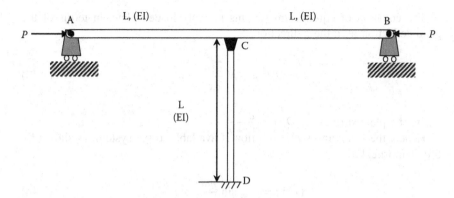

FIGURE 7.7   The three-column system with fixed support.

The final buckling deflection of this system is shown in Figure 7.8 below.

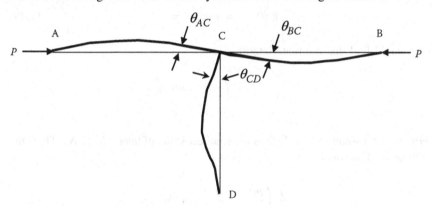

FIGURE 7.8   The three-column system buckling deformation.

7.5   BUCKLING OF PLATES

Consider a plate as shown by Figure 7.9, with thickness, t; and the length and width dimensions of a and b, respectively. Also, assume the plate is simply supported along the edges at $x = 0$ and $x = a$. A loading force, N_x, per unit length is applied to the edges of the plate laterally.

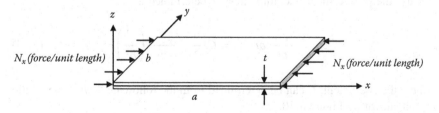

FIGURE 7.9   The simply supported plate buckling.

The equation of equilibrium for this laterally loaded plate, in terms of the plate deflection w in the z direction, is

$$D\nabla^4 w + N_x \frac{\partial^2 w}{\partial x^2} = 0 \qquad (7.47)$$

where the plate stiffness is $D = \frac{Et^3}{12(1-v^2)}$.

Hence, the following series solution is available to the system, as shown by S.P. Timoshenko,

$$w(x, y) = A_{mn} \sin \frac{m\pi x}{a} \sin \frac{n\pi y}{b} \qquad (7.48)$$

Applying the applicable boundary conditions,

$$w(0, y) = w(a, y) = 0 \qquad (7.49)$$

one would find that the plate lateral load is

$$N_x = \frac{\pi^2 D}{b^2} \left(\frac{mb}{a} + \frac{n^2 a}{mb} \right)^2 \qquad (7.50)$$

For buckling solution, $n = 1$ gives the smallest value of lateral load, N_x. Thus, the critical buckling load is

$$N_{x,cr} = \frac{\pi^2 D}{b^2} \left(\frac{mb}{a} + \frac{a}{mb} \right)^2 \quad \text{or} \quad N_{x,cr} = C \frac{\pi^2 D}{b^2} \qquad (7.51)$$

where $C = \left(\frac{mb}{a} + \frac{a}{mb} \right)^2$.

Applying the stiffness, D value, one would have the plate critical loading of

$$N_{x,cr} = C \frac{\pi^2 Et^3}{12(1 - v^2)b^2} \qquad (7.52)$$

Lastly, the plate critical buckling stress is determined as

$$\sigma_{cr} = \frac{N_{x,cr}}{t} \quad \text{or} \quad \sigma_{cr} = C \frac{\pi^2 E}{12(1 - v^2)} \left(\frac{t^2}{b^2} \right) \qquad (7.53)$$

where the coefficient, C can be determined from the following "a/b" aspect ratio graph shown by Figure 7.10.

To use the graph, simply calculate the aspect ratio a/b, then find the corresponding value based on a straight line to the lower bonds of the curves drawn. For example, for a/b aspect ratios are 1, 2, 3, 4 and 5 and the buckling coefficients are 4 in all cases.

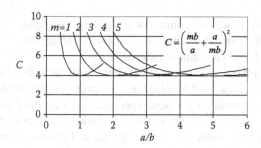

FIGURE 7.10   The plate buckling load coefficient versus the aspect ratio.

Example 7.2: For a simply supported plate of 20 in. long and 10 in. wide, with thickness of 0.01 in., determine the critical buckling stress. Assume a steel plate.

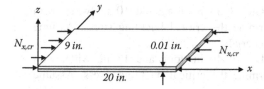

The plate aspect ratios would be

$$\frac{a}{b} = \frac{20}{9} = 2.22,$$

From Figure 7.10, the corresponding plate coefficient value, C, is determined as C = 4.036.

For the material properties of

$$E = 29 \times 10^6 \, psi$$
$$v = .3$$

The buckling stress is determined by expression (7.53) as

$$\sigma_{cr} = C \frac{\pi^2 E}{12(1 - v^2)} \left(\frac{t^2}{b^2} \right)$$

$$\sigma_{cr} = (4.036) \frac{\pi^2 (29 \times 10^6)}{12(1 - 0.3^2)} \left(\frac{0.01^2}{9^2} \right) = 130.6 \, psi$$

Problems

1. For a simply supported column of 10 ft. long with square cross-section of 2 in. by 2 in., determine the critical buckling stress. Assume an aluminum 6061-T6 column material.

2. For a fixed-fixed column of 20 ft. long with an I cross-section, determine the critical buckling load and the stress for the column. Assume the column is made of stainless steel material, with the flange width of 6 in., web height of 4.5 in. and 0.5 in. thickness all around.

3. For a fixed-free column of 15 ft. long with an I cross-section, determine the critical buckling load and the stress for the column. Assume the column is made of aluminum 7075 material, with the flange width of 5 in., web height of 4 in. and 0.5 in. thickness all around.

4. Consider a fixed-fixed column with a square cross-section of 3 in. by 3 in., with a desired minimum critical load of 1000 lb. Determine the maximum length. Assume the column is made of aluminum 7075-T6 material.

5. For a simply supported plate of 35 in. long and 10 in. wide, with thickness of 0.04 in., determine the critical buckling stress. Assume aluminum 7075-T6 plate material.

6. For a simply supported plate of 40 in. long and 13 in. wide, with thickness of 0.05 in., determine the critical buckling stress. Assume annealed steel plate material.

REFERENCES

Simitses, G., *An Introduction to Elastic Stability of Structures*, 1976. New York: Prentice Hall.

Timoshenko, S.P., Woinowsky-Krieger, S., *Theory of Plates and Shells*, 1959. New York: McGraw Hill.

8 Energy Methods

8.1 INTRODUCTION

This chapter explores the application of energy methods in determination of structural behavior for determinate and indeterminate structures. The strain energy is examined for determination of displacements and rotations in solid elastic bodies. Furthermore, potential energy applications for determination of the stresses due to impact loading on elastic bodies is investigated.

8.2 STRAIN ENERGY

Strain energy is defined as the energy stored within an elastic body when solid is deformed by an applied loading. Thus, the strain energy is equivalent to the work done by the external forces acting on a body. The general expression defining the strain energy is

$$U = \frac{1}{2} \int_V (\sigma_{xx}\varepsilon_{xx} + \sigma_{yy}\varepsilon_{yy} + \sigma_{zz}\varepsilon_{zz} + \tau_{xy}\gamma_{xy} + \tau_{yz}\gamma_{yz} + \tau_{xz}\gamma_{xz})dV \qquad (8.1)$$

For a bar of length L under only the axial loading, F, the expression above reduces to

$$U = \frac{1}{2} \int_V (\sigma_{xx}\varepsilon_{xx})dV \qquad (8.2)$$

where the volume change is $dV = Adx$, the area times the change in length.

After substituting the terms for the axial stress and strain, one would have

$$U = \frac{1}{2} \int_L \left(\frac{F}{A}\frac{F}{EA}\right)Adx \qquad (8.3)$$

This expression is solved and reduced as

$$U = \frac{F^2 L}{2EA} \qquad (8.4)$$

which for a series of n bars joined together, the strain energy takes the series form

DOI: 10.1201/9781003311218-8

$$U = \sum_{i=1}^{n} \frac{F_i^2 L_i}{2E_i A_i} \tag{8.5}$$

Likewise, for a beam under bending moment M, the strain energy expression reduces to

$$U = \frac{1}{2} \int_V (\sigma_{xx} \varepsilon_{xx}) dV \tag{8.6}$$

where the volume change is $dV = dAdx$, the change in area times the change in length.

After substituting the terms for the bending stress and strain, one would have

$$U = \frac{1}{2} \int_L \int_A \left(\frac{My}{I} \frac{My}{EI} \right) dAdx \tag{8.7}$$

where $I = \int_A y^2 dA$; this expression is solved and reduced as

$$U = \frac{1}{2} \int_L \frac{M^2}{EI} dx \tag{8.8}$$

For a beam under torque T, the strain energy expression reduces to

$$U = \frac{1}{2} \int_V (\tau_{xy} \gamma_{xy}) dV \tag{8.9}$$

where the volume change is $dV = dAdx$, the change in area times the change in length.

After substituting the terms for the shearing stress and strain, one would have

$$U = \frac{1}{2} \int_L \int_A \frac{Tr}{J} \frac{Tr}{JG} dAdx \tag{8.10}$$

where $J = \int_A r^2 dA$; this expression is solved and reduced as

$$U = \frac{1}{2} \int_L \frac{T^2}{JG} dx \tag{8.11}$$

which can be integrated, and the strain energy is

$$U = \frac{T^2}{2JG} \tag{8.12}$$

For a series of n beams joined together, the strain energy takes the series form

$$U = \sum_{i=1}^{n} \frac{T_i^2}{2J_i G_i}$$ (8.13)

8.3 CASTIGLIANO'S THEORY

Castigliano's theory states that in an elastic body that is under applied loading Q, the deflection, y_Q, at the point of the application of the load Q is the partial derivative of the strain energy of the structure, U, with respect to the load Q.

$$y_Q = \frac{\partial U}{\partial Q}$$ (8.14)

Likewise, by this theorem the slope of a beam, θ, where a moment, M, is applied is determined as the partial derivative of the strain energy, U, with respect to the applied moment, M.

$$\theta = \frac{\partial U}{\partial M}$$ (8.15)

Also the angle of the twist, ϕ, for a shaft where a torque, T, is applied is the partial derivative of the strain energy, U, with respect to the torque, T.

$$\phi = \frac{\partial U}{\partial T}$$ (8.16)

It should be noted that at a point where deflection is needed and there is no applied loading at that point, a fictitious load would be added and the partial derivative is carried out. Then, the fictitious load is set to null to determine the deflection term.

Example 8.1: The aluminum truss system shown is under an applied loading Q. The rods have circular cross-sections with 2 in. diameters. Determine the vertical displacement at point C where the loading is applied by using the energy method.

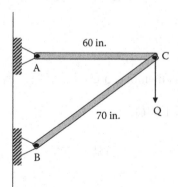

By fundamentals of static, the loads in the truss members are

$$F_{AC} = -1.67Q \text{ and } F_{BC} = 1.94Q$$

The cross-sectional area of the rods are: $A = \frac{\pi 2^2}{4} = 3.1415 in^2$
  Thus, the strain energy in the system by expression (8.5) is

$$U = \sum_{i=1}^{2} \frac{F_i^2 L_i}{2E_i A_i}$$

$$U = \frac{(1.67Q)^2 (60)}{2(10.6 \times 10^6)(3.1415)} + \frac{(1.94Q)^2 (70)}{2(10.6 \times 10^6)(3.1415)} = 0.647 \times 10^{-5}Q^2$$

$$y_c = \frac{dU}{dF} = 2(0.647 \times 10^{-5})Q \text{ thus } y_c = 0.13 \times 10^{-4}(Q) \text{ in.}$$

Example 8.1: (S.I. Units)

The aluminum truss system shown is under an applied loading Q. The rods have circular cross-sections with 0.051 m diameters. Determine the vertical displacement at point C, where the loading is applied by using the energy method.

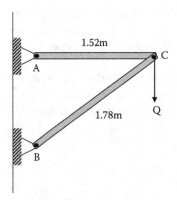

By fundamentals of static, the loads in the truss members are

$$F_{AC} = -1.67Q \text{ and } F_{BC} = 1.94Q$$

The cross-sectional area of the rods are: $A = \frac{\pi (.051)^2}{4} = 2.043E - 3 m^2$

Thus, the strain energy in the system by expression (8.5) is

$$U = \sum_{i=1}^{2} \frac{F_i^2 L_i}{2E_i A_i}$$

$$U = \frac{(1.67Q)^2(1.52)}{2(7.31 \times 10^{10})(2.043E - 3)} + \frac{(1.94Q)^2(1.78)}{2(7.31 \times 10^{10})(2.043E - 3)}$$

$$= 3.662 \times 10^{-8}Q^2$$

$$y_c = \frac{dU}{dF} = 2(3.662 \times 10^{-8})Q \text{ thus } y_c = 7.324 \times 10^{-8}(Q)m$$

Example 8.2: For the uniformly distributed loaded beam shown, determine the end deflection at point B by use of the energy method.

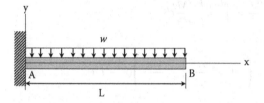

By Castigliano's method, a fictitious vertical load Q is introduced to point B as shown.

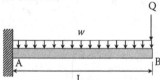

The free-body diagram of the system is developed that includes this loading.

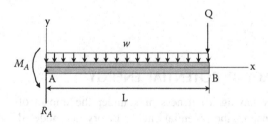

From mechanics of material, the reaction shear and moment at point A are determined as

$$R_A = Q + wL$$

$$M_A = QL + \frac{wL^2}{2}$$

The moment along the beam length at any point x is represented by

$$M(x) = (Q + wL)x - \left(QL + \frac{wL^2}{2}\right) - \frac{wx^2}{2}$$

The deflection at point B by Castigliano's theory is the partial derivative of expression (8.8) with respect to load Q, as stated by expression (8.14).

$$y_B = \frac{dU}{dQ} = \int_0^L \frac{M}{EI} \frac{dM}{dQ} dx$$

$$y_B = \int_0^L \left((Q + wL)x - \left(QL + \frac{wL^2}{2}\right) - \frac{wx^2}{2}\right)(x - L)\,dx$$

$$y_B = \frac{1}{EI}\left((Q + wL)\left(\frac{x^3}{3} - \frac{x^2L}{2}\right) - \left(QL + \frac{wL^2}{2}\right)\left(\frac{x^2}{2} - Lx\right) - \frac{wx^4}{8} + \frac{wx^3L}{6}\right)\Bigg|_{Q=0}^{x=L}\Bigg|_{x=0}$$

$$y_B = \frac{1}{EI}\left(\frac{wL^4}{3} - \frac{wL^4}{2} - \frac{wL^4}{4} + \frac{wL^4}{2} - \frac{wL^4}{8} + \frac{wL^4}{6}\right)$$

The deflection at point B is

$$y_B = \frac{wL^4}{8EI}$$

8.4   STRESS DUE TO IMPACT BY POTENTIAL ENERGY

The response of an elastic body having a stiffness of k under the impact of a weight, W, can be shown by use of the potential energy theory, as modeled by Figure 8.1. The potential energy of the falling mass is equal to the energy absorbed by the elastic body

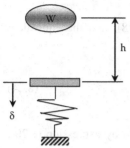

FIGURE 8.1   The impact loading of a elastic body.

$$W(h + \delta) = \frac{1}{2}F_{spring}\delta \tag{8.17}$$

where F_{spring} is the equivalent static load that the elastic body (spring) would produce.

This load is defined as

$$F_{spring} = k\delta \tag{8.18}$$

Now, substituting expression (8.18) into (8.17), one would have

$$W(h + \delta) = \frac{1}{2}k\delta^2 \tag{8.19}$$

The stiffness of the spring can be rewritten in terms of the static deflection δ_{st} of the spring:

$$W = k\delta_{st} \text{ or } k = \frac{W}{\delta_{st}} \tag{8.20}$$

By substitution of this stiffness term into expression (8.19), one would have

$$W(h + \delta) = \frac{1}{2}\frac{W}{\delta_{st}}\delta^2 \tag{8.21}$$

Re-writing the equation into form,

$$\frac{1}{2}\frac{W}{\delta_{st}}\delta^2 + W\delta + Wh = 0 \tag{8.22}$$

then, the solution for this system is

$$\delta = \delta_{st}\left(1 + \sqrt{1 + \frac{2h}{\delta_{st}}}\right) \tag{8.23}$$

Substitute this deflection equation into expression (8.18), and the equivalent spring force is derived as

$$F_{spring} = k\delta_{st}\left(1 + \sqrt{1 + \frac{2h}{\delta_{st}}}\right)$$ (8.24)

The term $k\delta_{st}$ can be replaced by the weight, W, as given by expression (8.20); thus,

$$F_{spring} = W\left(1 + \sqrt{1 + \frac{2h}{\delta_{st}}}\right)$$ (8.25)

This is the equivalent spring force due to the impact of the weight, in terms of the displacements. By fundamentals of physics, this relation can be re-written in terms of the impact velocity by using the relation

$$v^2 = 2gh$$ (8.26)

as

$$F_{spring} = W\left(1 + \sqrt{1 + \frac{v^2}{\delta_{st}g}}\right)$$ (8.27)

Now, consider a cantilever beam of length L while being impacted by a free falling weight, W, with the velocity of v (Figure 8.2).

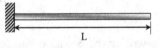

FIGURE 8.2 The cantilever beam under impact loading.

The stress due to this impact can be evaluated by determining the equivalent force on the beam due to the free falling weight by expression (8.27) as

$$Fe = W\left(1 + \sqrt{1 + \frac{v^2}{\delta_{st}g}}\right)$$ (8.28)

By mechanics of material, the maximum static deflection of the beam is

$$\delta_{st} = \frac{WL^3}{3EI} \tag{8.29}$$

Substitute back this displacement into expression (8.28), and one would have

$$Fe = W\left(1 + \sqrt{1 + \frac{3EIv^2}{WgL^3}}\right) \tag{8.30}$$

The resulting bending moment on the beam is

$$M = FeL \quad \text{or} \quad M = WL\left(1 + \sqrt{1 + \frac{3EIv^2}{WgL^3}}\right) \tag{8.31}$$

Thus, the stress on the beam due to the impact of the falling weight would be

$$\sigma = \frac{Mc}{I} = \frac{WL\left(1 + \sqrt{1 + \frac{3EIv^2}{WgL^3}}\right)c}{I} \tag{8.32}$$

Example 8.3: Consider a beam supported vertically by two spring systems, as shown by the following figure. A weight of 100 lb is dropped on the beam center from a distance of 10 in. high. Determine the maximum stress of the beam due to this impact loading by application of potential energy. The beam is made of aluminum ($E = 10 \times 10^6 psi$).

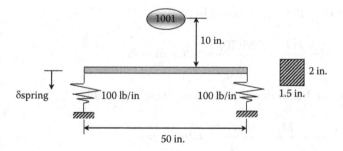

By expression (8.25), the equivalent force by the falling weight is

$$Fe = W\left(1 + \sqrt{1 + \frac{2h}{\delta_{st}}}\right) \quad Fe = 100\left(1 + \sqrt{1 + \frac{2(10)}{\delta_{st}}}\right)$$

The static equilibrium deflection is the sum of the beam deflection and the spring displacement.

The beam displacement is

$$\delta_{beam} = \frac{WL^3}{48EI} = \frac{100(50)^3}{48(10 \times 10^6)(1)} = 0.026 \ in.$$

The spring displacement is

$$\delta_{spring} = \frac{1}{2}\left(\frac{W}{k}\right) = \frac{1}{2}\left(\frac{100}{100}\right) = 0.5 \ in.$$

The total static equilibrium deflection is

$$\delta_{st} = \delta_{beam} + \delta_{spring} \quad \delta_{st} = 0.026 + 0.5 = 0.526 \ in.$$

Thus,

$$Fe = 100\left(1 + \sqrt{1 + \frac{2(10)}{0.526}}\right) = 725 \ lb$$

Based on the free-body diagram of the system and the fundamentals of the mechanics of material,

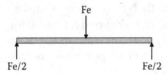

the bending moment of the beam would be

$$M = \frac{FeL}{4} = \frac{725(50)}{4} = 9063 \ in - lb$$

Thus, the stress on the beam due to this falling weight would be

$$\sigma = \frac{Mc}{I} = \frac{9063(1)}{1} = 9063 \ psi$$

Example 8.3: (S.I. Unit)

Consider a beam supported vertically by two spring systems, as shown by the following figure. A weight of 100 lb is dropped on the beam center from a distance of 10 in. high. Determine the maximum stress of the beam due to this impact loading by application of potential energy. The beam is made of aluminum ($E = 10 \times 10^6 psi$).

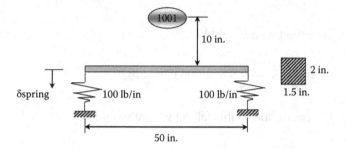

By expression (8.25), the equivalent force by the falling weight is

$$Fe = W\left(1 + \sqrt{1 + \frac{2h}{\delta_{st}}}\right) \quad Fe = 100\left(1 + \sqrt{1 + \frac{2(10)}{\delta_{st}}}\right)$$

The static equilibrium deflection is the sum of the beam deflection and the spring displacement.

The beam displacement is

$$\delta_{beam} = \frac{WL^3}{48EI} = \frac{100(50)^3}{48(10 \times 10^6)(1)} = 0.026 \ in.$$

The spring displacement is

$$\delta_{spring} = \frac{1}{2}\left(\frac{W}{k}\right) = \frac{1}{2}\left(\frac{100}{100}\right) = 0.5 \ in.$$

The total static equilibrium deflection is

$$\delta_{st} = \delta_{beam} + \delta_{spring} \quad \delta_{st} = 0.026 + 0.5 = 0.526 \ in.$$

Thus,

$$Fe = 100\left(1 + \sqrt{1 + \frac{2(10)}{0.526}}\right) = 725 \ lb$$

Based on the free-body diagram of the system and the fundamentals of the mechanics of material,

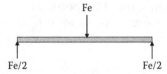

the bending moment of the beam would be

$$M = \frac{FeL}{4} = \frac{725(50)}{4} = 9063 \ in - lb$$

Thus, the stress on the beam due to this falling weight would be

$$\sigma = \frac{Mc}{I} = \frac{9063(1)}{1} = 9063 \ psi$$

Problems

1. A 80 lb weight is dropped in the middle of a simply supported beam from a height of 30 in. The simply supported beam is 70 in. long with a square cross-section of 1.2 in. by 1.2 in. The beam is made of annealed steel material. Determine the maximum stress and the deflection of the beam due to this impact.
2. Consider an aluminum 7075 cantilevered beam of 50 in. long with a rectangular cross-section of 1.5 in. wide by 2 in. high. If a weight of 50 lb is dropped onto the free-end of the beam, what is the free fall height that causes the beam to permanently deform?
3. For the beam system shown, determine the deflection at point A using the Castigliano's theorem. (Assume E and I properties.)

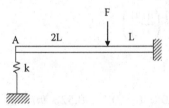

4. For the beam system shown, determine the deflection at point E. (Assume E and I properties.)

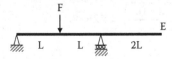

5. For the truss system shown, determine the deflection at point C. Assume all members are made from round 1 in. diameter bars. The bar material is steel ($E = 29 \times 10^6$ psi).

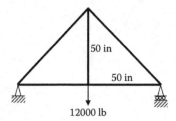

50 in

50 in

12000 lb

6. For the truss system shown, determine the deflection at the point of the load application. Assume all members are made from round 0.5 in. diameter bars. The bar material is cold rolled steel.

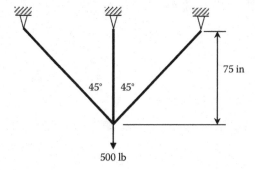

75 in

45°   45°

500 lb

REFERENCES

Megson, T.H.G., *Structural and Stress Analysis*, 1996. New York: Halsted Press.

Ugural, A.C., Fenster, S.K., *Advanced Strength and Applied Elasticity*, 1995. New Jersey: Prentice Hall.

REFERENCE

9 Fatigue Analysis

9.1 INTRODUCTION

Fatigue is known as the failure of the structural component under repeated loading. It can occur prematurely and reduce the life of a component considerably. There are three methods currently available for fatigue analysis. They are: 1) the stress-life approach, 2) the strain-life approach and 3) the fracture mechanics approach. This chapter only briefly concentrates on the stress-life approach. It is written to give the stress analyst a general idea of stress-life fatigue analysis. For a detailed fatigue analysis technique, refer to a metal fatigue text or handbook.

9.2 STRESS-LIFE S-N CURVE

The plot of the alternating stress (S) versus the cycles-to-failure (N) is known as the S-N curve, which is used to predict the material failure under repeated loads. In the case of steel, this curve can be quantified by the expression shown by the relation below:

$$S = 10^C N^b \text{ (for } 10^3 < N < 10^6) \tag{9.1}$$

where $b = -\frac{1}{3}\log_{10}\frac{0.9S_{ult}}{0.5S_{ult}}$ and $C = \log_{10}\frac{(0.9S_{ult})^2}{0.5S_{ult}}$.

The graphical representation of expression (9.1) is shown in Figure 9.1 below in log scale.

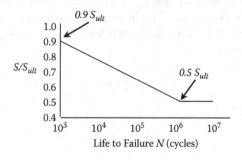

FIGURE 9.1 The general S-N curve for steel.

DOI: 10.1201/9781003311218-9

The stress level at the life of 1 million cycles is referred to as the endurance limit (S_{end}) of the material. In Figure 9.1, the endurance limit of the material is where the alternating stress flattens down ($S_{end} = 0.5\ S_u$). In reality, for fatigue analysis this endurance limit is factored out for size, type of loading, temperature, environment, surface treatment and surface finish as applicable. These factors are available for each material in their respective material's handbook.

In fatigue analysis, the mean stress, stress amplitude and stress ratio are utilized often. The mean stress is defined as

$$\sigma_{mean} = \frac{\sigma_{max} + \sigma_{min}}{2} \tag{9.2}$$

where σ_{max} is the maximum stress level the component undergoes over time and σ_{min} is the minimum stress level the component undergoes over time.

Likewise, the stress amplitude is defined as

$$\sigma_{amp} = \frac{\sigma_{max} - \sigma_{min}}{2} \tag{9.3}$$

and stress ratio, or R, is defined as

$$R = \frac{\sigma_{min}}{\sigma_{max}} \tag{9.4}$$

In 1899, Goodman developed the fatigue relationship,

$$\frac{\sigma_{amp}}{S_{end}} + \frac{\sigma_{mean}}{S_u} = 1 \tag{9.5}$$

which can be used to derive an expression for the fully reversed stress level (S_n), as shown by expressions (9.6) and (9.7). This stress level is used to predict the fatigue life of a component under cyclic stress. The corresponding life from S-N curve or expression (9.1) is the total predicted life of the component under repeated cyclic loading:

$$\frac{\sigma_{amp}}{S_n} + \frac{\sigma_{mean}}{S_u} = 1 \tag{9.6}$$

$$S_n = \frac{\sigma_{amp}}{1 - \frac{\sigma_{mean}}{S_u}} \tag{9.7}$$

Example 9.1: Consider a structural component that undergoes cyclic stress with maximum level of 100 ksi and minimum level of 2 ksi. The component is made out of steel with ultimate strength of 125 ksi. Determine the life of the component under this condition.

$$\sigma_{max} = 100 \text{ ksi}$$

and

$$\sigma_{min} = 2 \text{ ksi}$$

$$\sigma_{mean} = \frac{\sigma_{max} + \sigma_{min}}{2}, \quad \sigma_{mean} = \frac{100 + 2}{2} = 51 \text{ ksi}$$

$$\sigma_{amp} = \frac{\sigma_{max} - \sigma_{min}}{2}, \quad \sigma_{amp} = \frac{100 - 2}{2} = 49 \text{ ksi}$$

Determine the fully reversed stress level using expression (9.6):

$$\frac{\sigma_{amp}}{S_n} + \frac{\sigma_{mean}}{S_u} = 1, \quad \frac{49}{S_n} + \frac{51}{125} = 1$$

Thus, $S_n = 82.9$ ksi.

Construct the S-N curve and determine the life, graphically, based on an alternating stress level of 82.9 ksi from that curve or use expression (9.1).

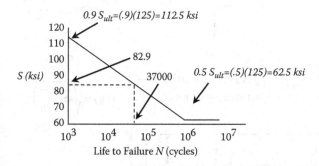

Thus, the predicted life of the component is N = 37,000 cycles.

9.3   FATIGUE CRACK GROWTH

The determination of the life of a structural component with the presence of a crack is examined here for fatigue analysis. Fracture mechanics principals are used to predict the life of a structure before an existing crack on the structure is grown to a critical detectable size.

In the 1960s, Paris developed an expression relating the crack growth rate to the stress cycles. This expression, as shown below (9.8), relates the crack length (a) rate and the fatigue cycles (N) to the stress intensity of the component under cyclic loading

$$\frac{da}{dN} = C(\Delta K)^m \qquad (9.8)$$

where the stress intensity $\Delta K = f(g)(\sigma_{max} - \sigma_{min})\sqrt{\pi a}$; "$C$" and "$m$" are material constants that can be obtained from fatigue material handbooks. Also, $f(g)$ is a function of the crack geometry and is defined by Bannantine et al. (1990), as shown in Figure 9.2.

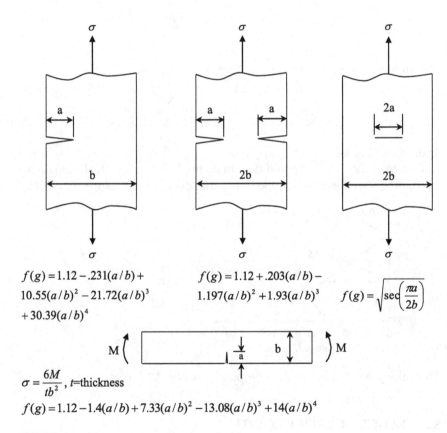

$f(g) = 1.12 - .231(a/b) + 10.55(a/b)^2 - 21.72(a/b)^3 + 30.39(a/b)^4$

$f(g) = 1.12 + .203(a/b) - 1.197(a/b)^2 + 1.93(a/b)^3$

$f(g) = \sqrt{\sec\left(\frac{\pi a}{2b}\right)}$

$\sigma = \dfrac{6M}{tb^2}$, t=thickness

$f(g) = 1.12 - 1.4(a/b) + 7.33(a/b)^2 - 13.08(a/b)^3 + 14(a/b)^4$

FIGURE 9.2   The stress intensity correction factors.

Now, by the separation of the variables of expression (9.8) and integration, the life-cycles-to-failure ($N_{failure}$) for the structural component can be calculated as

$$\int dN = \int \frac{da}{C(\Delta K)^m} \qquad (9.9)$$

or

$$N_{failure} = \int_{ai}^{af} \frac{da}{C(\Delta K)^m} \qquad (9.10)$$

where "ai" is the existing initial crack size and "af" is the final critical crack size. This final crack size is usually calculated as

$$af = \frac{1}{\pi}\left(\frac{Kc}{f(g)\sigma_{max}}\right)^2 \qquad (9.11)$$

with "Kc" being the critical stress intensity.

Example 9.2: Consider a structural component that undergoes a cyclic bending with maximum to minimum level of $0 \leq M \leq 1200$ lbf-in. The component is made out of steel with a critical stress intensity of $Kc = 73$ ksi (in.)$^{1/2}$. The material constant $C = 3.8 \times 10^{-11}$ and $m = 3.0$. The initial crack noticed on the member is 0.004 in. Determine the failure life cycles of the component under this condition ($b = 0.5$ in. and $t = 0.25$ in.).

Find the change in stress intensity by

$$\Delta K = f(g)(\sigma_{max} - \sigma_{min})\sqrt{\pi a}$$

where maximum and minimum stress can be calculated by the equation $\sigma = \frac{6M}{tb^2}$ and $f(g) = 1.12 - 1.4(a/b) + 7.33(a/b)^2 - 13.08(a/b)^3 + 14(a/b)^4$.

$$\sigma_{max} = 115.2 \ ksi$$

$$\sigma_{min} = 0 \ ksi$$

$$f(g) = 1.12 - 1.4(.004/.5) + 7.33(.004/.5)^2 - 13.08(.004/.5)^3$$
$$+ 14(.004/.5)^4 = 1.11$$

and

$$\Delta K = (1.11)(115.2 - 0)\sqrt{\pi a}$$

Finding the critical crack size by using equation $af = \dfrac{1}{\pi}\left(\dfrac{Kc}{f(g)\sigma_{max}}\right)^2$

$$af = \frac{1}{\pi}\left(\frac{73}{(1.11)(115.2)}\right)^2, \quad af = 0.1037\ in.$$

Finally, use equation $N_{failure} = \int_{ai}^{af} \dfrac{da}{C(\Delta K)^m}$ to find the number of cycles to failure.

$$N_{failure} = \int_{0.004}^{0.1037} \frac{da}{3.8 \times 10^{-11}((1.11)(115.2 - 0)\sqrt{\pi a})^3}$$

$$N_{failure} = 57000\ cycles$$

Problems

1. A structural component undergoes a cyclic stress with maximum level of 65 ksi and minimum level of 12 ksi. The component is made out of steel with ultimate strength of 70 ksi. Determine the life of the component under this condition.
2. A structural component undergoes a cyclic stress with maximum level of 70 ksi and minimum level of 30 ksi. The component is made out of steel with ultimate strength of 95 ksi. Determine the life of the component under this condition.
3. A relatively large plate contains a center crack of 0.02 inches wide. The yield strength of the material is 70 ksi, and the fracture toughness of the material is 110 ksi $(in)^{1/2}$. The plate is subjected to a maximum stress level of 45 ksi and a minimum stress level of 5 ksi. Assuming the fatigue material constants $C = 10^{-8}$ and $m = 3.5$, determine the number of cycles to failure for this plate.

REFERENCE

Bannantine, J.A., Comer, J.J., Handrock, J.L., *Fundamental of Metal Fatigue Analysis*, 1990. New Jersey: Prentice Hall.

10 Numerical and Finite Element Methods

10.1 INTRODUCTION

The finite element method is one of the commonly used methods for calculation of the stresses and deflections of the large truss and beam systems. The structural system is discretely divided into finite elements that each have their own basic equilibrium models. Each element model is defined and assembled into the larger global model that defines the system. Then, the system is represented by simultaneous equations and solved by numerical methods. Hence, the nodal displacements are determined, and the individual element forces are computed. The displacements can be translated into strains, and the forces can be translated into stress values for each element.

10.2 STRESS ON TRUSS ELEMENTS

By definition, a truss is a structural system of slender, "two-force" members or rods joined with pins and loaded only at the joints. Consider a rod with an original length of L and applied force F at the end joints.

FIGURE 10.1 The truss (rod) member under axial loading.

Under applied loading, the rod elongates an amount of horizontal displacement, u. The strain on the rod is defined as

$$\varepsilon = \frac{u}{L} \tag{10.1}$$

and the axial stress is defined as

$$\sigma = \frac{F}{L} \tag{10.2}$$

By Hooke's law, the stress and strain are related as such:

$$\sigma = E\varepsilon \tag{10.3}$$

DOI: 10.1201/9781003311218-10

Thus, by substitution of (10.1) and (10.2) into the Hooke's relation, one would have

$$\frac{F}{A} = E\frac{u}{L} \qquad (10.4)$$

or for displacement,

$$u = \frac{FL}{AE} \qquad (10.5)$$

Solving for force as a function of displacement, one would have

$$F = \frac{AE}{L}u \qquad (10.6)$$

where $\frac{AE}{L}$ is the stiffness of the rod. Let k represent this stiffness:

$$k = \frac{AE}{L} \qquad (10.7)$$

Then, expression (10.6) can be rewritten as

$$F = ku \qquad (10.8)$$

Now, consider the rod under equilibrium conditions. The summation of forces on the rod must equal zero (Figure 10.2).

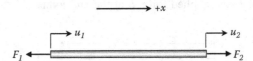

FIGURE 10.2 The rod, "truss" member under equilibrium conditions.

$$\sum F = 0, \;\; F_1 + F_2 = 0 \text{ or } F_1 = -F_2 \qquad (10.9)$$

The same concept is applicable to the rod elongation. Thus, total rod elongation can be represented by the individual end displacements:

$$u = u_2 - u_1 \qquad (10.10)$$

Apply expression (10.8) and the force on the rod is

$$F_2 = -F_1 = k(u_2 - u_1) \qquad (10.11)$$

or

$$F_1 = u_1 k - u_2 k \qquad (10.12)$$

and

$$F_2 = u_2 k - u_1 k \qquad (10.13)$$

Expressions (10.12) and (10.13) can be represented in the matrix form as

$$\left\{ \begin{array}{c} F_1 \\ F_2 \end{array} \right\} = \left[\begin{array}{cc} k & -k \\ -k & k \end{array} \right] \left\{ \begin{array}{c} u_1 \\ u_2 \end{array} \right\} \qquad (10.14)$$

which follows the form of

$$\{R\} = [K]\{D\} \qquad (10.15)$$

where $\{R\}$ is the load vector, $\{D\}$ is the displacement vector and $[K]$ is the stiffness matrix.

$$[K] = \left[\begin{array}{cc} k & -k \\ -k & k \end{array} \right] \qquad (10.16)$$

This is the finite element representation of the rod, "truss" member.

Let's consider the same element except at an inclined angle of θ. The load and the end displacements can be shown as in Figure 10.3, below.

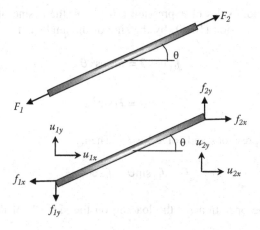

FIGURE 10.3 The rod, "truss" member at an inclined angle.

As shown in this figure, the loads and the end displacements can be resolved into their respective components in terms of the inclined angle, for side (1) as

$$\cos\theta = \frac{f_{1x}}{F_1} \qquad (10.17)$$

$$\sin\theta = \frac{f_{1y}}{F_1} \qquad (10.18)$$

Hence,

$$f_{1x} = F_1 \cos\theta \qquad (10.19)$$

$$f_{1y} = F_1 \sin\theta \qquad (10.20)$$

Likewise, for the side (2) of the rod

$$\cos\theta = \frac{f_{2x}}{F_2} \qquad (10.21)$$

$$\sin\theta = \frac{f_{2y}}{F_2} \qquad (10.22)$$

Hence,

$$f_{2x} = F_2 \cos\theta \qquad (10.23)$$

$$f_{2y} = F_2 \sin\theta \qquad (10.24)$$

Now, multiply both sides of expression (10.19) by the cosine of the angle and both sides of expression (10.20) by the sine of the angle as follows:

$$f_{1x} \cos\theta = F_1 \cos^2\theta \qquad (10.25)$$

$$f_{1y} \sin\theta = F_1 \sin^2\theta \qquad (10.26)$$

Add the two expressions and solve for F_1. Then,

$$F_1 = f_{1y} \sin\theta + f_{1x} \cos\theta \qquad (10.27)$$

Repeat the same operation for the loading on the side (2) of the rod and one would have

$$f_{2x} \cos\theta = F_2 \cos^2\theta \qquad (10.28)$$

$$f_{2y} \sin\theta = F_2 \sin^2\theta \qquad (10.29)$$

and hence,

$$F_2 = f_{2y} \sin\theta + f_{2x} \cos\theta \qquad (10.30)$$

By the same analogy, the end displacements can be solved as

$$u_1 = u_{1x} \cos\theta + u_{1y} \sin\theta \qquad (10.31)$$

and

$$u_2 = u_{2x} \cos\theta + u_{2y} \sin\theta \qquad (10.32)$$

where the total rod displacement is

$$u = u_2 - u_1 = u_{2x} \cos\theta + u_{2y} \sin\theta - u_{1x} \cos\theta - u_{1y} \sin\theta \qquad (10.33)$$

Now, by equilibrium,

$$F_1 = -k(u_2 - u_1) \qquad (10.34)$$

$$F_2 = k(u_2 - u_1) \qquad (10.35)$$

Substitute the displacement terms into these expressions and one would have the following expressions for the rod end loadings:

$$F_1 = ku_{1x} \cos\theta + ku_{1y} \sin\theta - ku_{2x} \cos\theta - ku_{2y} \sin\theta \qquad (10.36)$$

$$F_2 = -ku_{1x} \cos\theta - ku_{1y} \sin\theta + ku_{2x} \cos\theta + ku_{2y} \sin\theta \qquad (10.37)$$

Finally, the rod end-force components can be derived by substitution of expressions (10.36) and (10.37) back into equations (10.19), (10.20), (10.23) and (10.24), as shown below:

$$f_{1x} = (k\cos^2\theta)u_{1x} + (k\sin\theta\cos\theta)u_{1y} + (-k\cos^2\theta)u_{2x} + (-k\sin\theta\cos\theta)u_{2y} \qquad (10.38)$$

$$f_{1y} = (k\sin\theta\cos\theta)u_{1x} + (k\sin^2\theta)u_{1y} + (-k\cos\theta\sin\theta)u_{2x} + (-k\sin^2\theta)u_{2y} \qquad (10.39)$$

$$f_{2x} = (-k\cos^2\theta)u_{1x} + (-k\sin\theta\cos\theta)u_{1y} + (k\cos^2\theta)u_{2x} + (k\sin\theta\cos\theta)u_{2y} \qquad (10.40)$$

$$f_{2y} = (-k\sin\theta\cos\theta)u_{1x} + (-k\sin^2\theta)u_{1y} + (k\cos\theta\sin\theta)u_{2x} + (k\sin^2\theta)u_{2y} \qquad (10.41)$$

Rewriting these expression in the matrix form,

$$\{f\} = [K][u] \qquad (10.42)$$

one would have developed the finite element relation of the rod based on the inclined angle as

$$
\begin{Bmatrix} f_{1x} \\ f_{1y} \\ f_{2x} \\ f_{2y} \end{Bmatrix} = k \begin{bmatrix} c^2 & sc & -c^2 & -sc \\ sc & s^2 & -sc & -s^2 \\ -c^2 & -sc & c^2 & sc \\ -sc & -s^2 & sc & s^2 \end{bmatrix} \begin{Bmatrix} u_{1x} \\ u_{1y} \\ u_{2x} \\ u_{2y} \end{Bmatrix}
\tag{10.43}
$$

where $s = \sin\theta$ and $c = \cos\theta$.

Now, consider the truss numerical problem, as shown below. The truss system is under an applied loading of 450 lb at the right corner and pinned at the left side as shown. The rods are all made of AL 2024-T4, with cross-sectional area of 2.33 in^2 (Figure 10.4).

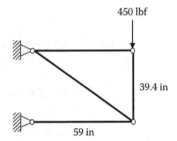

FIGURE 10.4   The truss system under loading.

The free-body diagram of the system is shown by Figure 10.5, where R_{ij} are the reaction loads.

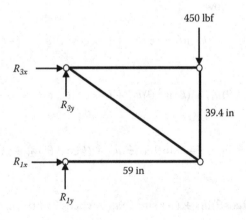

FIGURE 10.5   The truss system free-body diagram.

The finite element representation of the system is shown by the following figure,

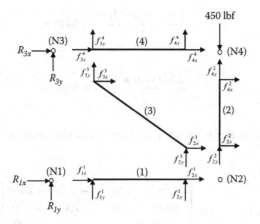

FIGURE 10.6 The truss system finite element model.

where each truss member is assigned the appropriate vertical and axial end-load components. Also, the external loads at the nodes are shown (Figure 10.6).

Table 10.1 below is the breakdown of the model.

TABLE 10.1
Truss System Data

Element	Nodes	Length	Area	θ	Sθ	Cθ
1	1,2	59 in.	2.33 in²	0°	0	1
2	2,4	39.4 in	2.33 in²	90°	1	0
3	2,3	71 in.	2.33 in²	146.26°	0.55	−0.83
4	3,4	59 in.	2.33 in²	0°	0	1

Each element stiffness is calculated as

$$k^{(e)} = \frac{A^{(e)} E^{(e)}}{L^{(e)}} \tag{10.44}$$

Hence, for elements (1) through (4), one would have

$$k^{(1)} = \frac{2.33(10.6 \times 10^6)}{59} = 418610 \tag{10.45}$$

$$k^{(2)} = \frac{2.33(10.6 \times 10^6)}{39.4} = 626853 \tag{10.46}$$

$$k^{(3)} = \frac{2.33(10.6 \times 10^6)}{71} = 347859 \tag{10.47}$$

$$k^{(4)} = \frac{2.33(10.6 \times 10^6)}{59} = 418610 \tag{10.48}$$

The stiffness matrices for each element are evaluated based on their respective inclined angles as

$$[K]^{(1)} = k^{(1)} \begin{bmatrix} c^2 0 & s0c0 & -c^2 0 & -s0c0 \\ s0c0 & s^2 0 & -s0c0 & -s^2 0 \\ -c^2 0 & -s0c0 & c^2 0 & s0c0 \\ -s0c0 & -s^2 0 & s0c0 & s^2 0 \end{bmatrix} \tag{10.49}$$

$$[K]^{(2)} = k^{(2)} \begin{bmatrix} c^2 90 & s90c90 & -c^2 90 & -s90c90 \\ s90c90 & s^2 90 & -s90c90 & -s^2 90 \\ -c^2 90 & -s90c90 & c^2 90 & s90c90 \\ -s90c90 & -s^2 90 & s90c90 & s^2 90 \end{bmatrix} \tag{10.50}$$

$$[K]^{(3)} = k^{(3)}$$

$$\begin{bmatrix} c^2 146.26 & s146.26c146.26 & -c^2 146.26 & -s146.26c146.26 \\ s146.26c146.26 & s^2 146.26 & -s146.26c146.26 & -s^2 146.26 \\ -c^2 146.26 & -s146.26c146.26 & c^2 146.26 & s146.26c146.26 \\ -s146.26c146.26 & -s^2 146.26 & s146.26c146.26 & s^2 146.26 \end{bmatrix} \tag{10.51}$$

$$[K]^{(4)} = k^{(4)} \begin{bmatrix} c^2 0 & s0c0 & -c^2 0 & -s0c0 \\ s0c0 & s^2 0 & -s0c0 & -s^2 0 \\ -c^2 0 & -s0c0 & c^2 0 & s0c0 \\ -s0c0 & -s^2 0 & s0c0 & s^2 0 \end{bmatrix} \tag{10.52}$$

Numerically, the stiffness matrices are

$$[K]^{(1)} = \begin{bmatrix} k_{11}^1 & k_{12}^1 & k_{13}^1 & k_{14}^1 \\ k_{21}^1 & k_{22}^1 & k_{23}^1 & k_{24}^1 \\ k_{31}^1 & k_{32}^1 & k_{33}^1 & k_{34}^1 \\ k_{41}^1 & k_{42}^1 & k_{43}^1 & k_{44}^1 \end{bmatrix} = \begin{bmatrix} 418610 & 0 & -418610 & 0 \\ 0 & 0 & 0 & 0 \\ -418610 & 0 & 418610 & 0 \\ 0 & 0 & 0 & 0 \end{bmatrix} \tag{10.53}$$

$$[K]^{(2)} = \begin{bmatrix} k^2_{11} & k^2_{12} & k^2_{13} & k^2_{14} \\ k^2_{21} & k^2_{22} & k^2_{23} & k^2_{24} \\ k^2_{31} & k^2_{32} & k^2_{33} & k^2_{34} \\ k^2_{41} & k^2_{42} & k^2_{43} & k^2_{44} \end{bmatrix} = \begin{bmatrix} 0 & 0 & 0 & 0 \\ 0 & 626853 & 0 & -626853 \\ 0 & 0 & 0 & 0 \\ 0 & -626853 & 0 & 626853 \end{bmatrix} \quad (10.54)$$

$$[K]^{(3)} = \begin{bmatrix} k^3_{11} & k^3_{12} & k^3_{13} & k^3_{14} \\ k^3_{21} & k^3_{22} & k^3_{23} & k^3_{24} \\ k^3_{31} & k^3_{32} & k^3_{33} & k^1_{34} \\ k^3_{41} & k^3_{42} & k^3_{43} & k^3_{44} \end{bmatrix} = \begin{bmatrix} 240546 & -160667 & -240546 & 160667 \\ -160667 & 107313 & 160667 & -107313 \\ -240546 & 160667 & 240546 & -160667 \\ 160667 & -107313 & -160667 & 107313 \end{bmatrix}$$

$$(10.55)$$

$$[K]^{(4)} = \begin{bmatrix} k^4_{11} & k^4_{12} & k^4_{13} & k^4_{14} \\ k^4_{21} & k^4_{22} & k^4_{23} & k^4_{24} \\ k^4_{31} & k^4_{32} & k^4_{33} & k^4_{34} \\ k^4_{41} & k^4_{42} & k^4_{43} & k^4_{44} \end{bmatrix} = \begin{bmatrix} 418610 & 0 & -418610 & 0 \\ 0 & 0 & 0 & 0 \\ -418610 & 0 & 418610 & 0 \\ 0 & 0 & 0 & 0 \end{bmatrix} \quad (10.56)$$

Knowing that the global nodal forces for N elements are

$$\{F\} = \sum_{e=1}^{N} \{f^{(e)}\} \quad (10.57)$$

one can sum up the loads at each node as

$$F_{1x} = f^1_{1x} \quad (10.58)$$

$$F_{1y} = f^1_{1y} \quad (10.59)$$

$$F_{2x} = f^1_{2x} + f^2_{2x} + f^3_{2x} \quad (10.60)$$

$$F_{2y} = f^1_{2y} + f^2_{2y} + f^3_{2y} \quad (10.61)$$

$$F_{3x} = f^3_{3x} + f^4_{3x} \quad (10.62)$$

$$F_{3y} = f^3_{3y} + f^4_{3y} \quad (10.63)$$

$$F_{4x} = f^2_{4x} + f^4_{4x} \quad (10.64)$$

$$F_{4y} = f^2_{4y} + f^4_{4y} \quad (10.65)$$

The nodal loads in vector form are

$$
\begin{Bmatrix} F_{1x} \\ F_{1y} \\ F_{2x} \\ F_{2y} \\ F_{3x} \\ F_{3y} \\ F_{4x} \\ F_{4y} \end{Bmatrix} = \begin{Bmatrix} f_{1x}^1 \\ f_{1y}^1 \\ f_{2x}^1 + f_{2x}^2 + f_{2x}^3 \\ f_{2y}^1 + f_{2y}^2 + f_{2y}^3 \\ f_{3x}^3 + f_{3x}^4 \\ f_{3y}^3 + f_{3y}^4 \\ f_{4x}^2 + f_{4x}^4 \\ f_{4y}^2 + f_{4y}^4 \end{Bmatrix} = \begin{Bmatrix} f_{1x}^1 \\ f_{1y}^1 \\ f_{2x}^1 \\ f_{2y}^1 \\ 0 \\ 0 \\ 0 \\ 0 \end{Bmatrix} + \begin{Bmatrix} 0 \\ 0 \\ f_{2x}^2 \\ f_{2y}^2 \\ 0 \\ 0 \\ f_{4x}^2 \\ f_{4y}^2 \end{Bmatrix} + \begin{Bmatrix} 0 \\ 0 \\ f_{2x}^3 \\ f_{2y}^3 \\ f_{3x}^3 \\ f_{3y}^3 \\ 0 \\ 0 \end{Bmatrix} + \begin{Bmatrix} 0 \\ 0 \\ 0 \\ 0 \\ f_{3x}^4 \\ f_{3y}^4 \\ f_{4x}^4 \\ f_{4y}^4 \end{Bmatrix}
$$

(1)   (2)   (3)   (4)

$$ \tag{10.66} $$

where the summation vector $\{F_{ij}\}$ can be broken down into force vectors for each element (1 through 4).

Now, the force vector for each element is the element stiffness times the displacement vector,

$$ \{f^{(e)}\} = [K]^{(e)} \{u\} \tag{10.67} $$

Hence, specifically for each element one would have

$$
\begin{Bmatrix} f_{1x}^1 \\ f_{1y}^1 \\ f_{2x}^1 \\ f_{2y}^1 \\ 0 \\ 0 \\ 0 \\ 0 \end{Bmatrix} = \begin{bmatrix} k_{11}^1 & k_{12}^1 & k_{13}^1 & k_{14}^1 & 0 & 0 & 0 & 0 \\ k_{21}^1 & k_{22}^1 & k_{23}^1 & k_{24}^1 & 0 & 0 & 0 & 0 \\ k_{31}^1 & k_{32}^1 & k_{33}^1 & k_{34}^1 & 0 & 0 & 0 & 0 \\ k_{41}^1 & k_{42}^1 & k_{43}^1 & k_{44}^1 & 0 & 0 & 0 & 0 \\ 0 & 0 & 0 & 0 & 0 & 0 & 0 & 0 \\ 0 & 0 & 0 & 0 & 0 & 0 & 0 & 0 \\ 0 & 0 & 0 & 0 & 0 & 0 & 0 & 0 \\ 0 & 0 & 0 & 0 & 0 & 0 & 0 & 0 \end{bmatrix} \begin{Bmatrix} u_{1x} \\ u_{1y} \\ u_{2x} \\ u_{2y} \\ 0 \\ 0 \\ 0 \\ 0 \end{Bmatrix}
\tag{10.68}
$$

$$
\begin{Bmatrix} 0 \\ 0 \\ f_{2x}^2 \\ f_{2y}^2 \\ 0 \\ 0 \\ f_{4x}^2 \\ f_{4y}^2 \end{Bmatrix} = \begin{bmatrix} 0 & 0 & 0 & 0 & 0 & 0 & 0 & 0 \\ 0 & 0 & 0 & 0 & 0 & 0 & 0 & 0 \\ 0 & 0 & k_{11}^2 & k_{12}^2 & 0 & 0 & k_{13}^2 & k_{14}^2 \\ 0 & 0 & k_{21}^2 & k_{22}^2 & 0 & 0 & k_{23}^2 & k_{24}^2 \\ 0 & 0 & 0 & 0 & 0 & 0 & 0 & 0 \\ 0 & 0 & 0 & 0 & 0 & 0 & 0 & 0 \\ 0 & 0 & k_{31}^2 & k_{32}^2 & 0 & 0 & k_{33}^2 & k_{34}^2 \\ 0 & 0 & k_{41}^2 & k_{42}^2 & 0 & 0 & k_{43}^2 & k_{44}^2 \end{bmatrix} \begin{Bmatrix} 0 \\ 0 \\ u_{2x} \\ u_{2y} \\ 0 \\ 0 \\ u_{4x} \\ u_{4y} \end{Bmatrix}
\tag{10.69}
$$

$$
\begin{Bmatrix} 0 \\ 0 \\ f_{2x}^3 \\ f_{2y}^3 \\ f_{3x}^3 \\ f_{3y}^3 \\ 0 \\ 0 \end{Bmatrix} = \begin{bmatrix} 0 & 0 & 0 & 0 & 0 & 0 & 0 & 0 \\ 0 & 0 & 0 & 0 & 0 & 0 & 0 & 0 \\ 0 & 0 & k_{11}^3 & k_{12}^3 & k_{13}^3 & k_{14}^3 & 0 & 0 \\ 0 & 0 & k_{21}^3 & k_{22}^3 & k_{23}^3 & k_{24}^3 & 0 & 0 \\ 0 & 0 & k_{31}^3 & k_{32}^3 & k_{33}^3 & k_{34}^3 & 0 & 0 \\ 0 & 0 & k_{41}^3 & k_{42}^3 & k_{43}^3 & k_{44}^3 & 0 & 0 \\ 0 & 0 & 0 & 0 & 0 & 0 & 0 & 0 \\ 0 & 0 & 0 & 0 & 0 & 0 & 0 & 0 \end{bmatrix} \begin{Bmatrix} 0 \\ 0 \\ u_{2x} \\ u_{2y} \\ u_{3x} \\ u_{3y} \\ 0 \\ 0 \end{Bmatrix} \tag{10.70}
$$

$$
\begin{Bmatrix} 0 \\ 0 \\ 0 \\ 0 \\ f_{3x}^4 \\ f_{3y}^4 \\ f_{4x}^4 \\ f_{4y}^4 \end{Bmatrix} = \begin{bmatrix} 0 & 0 & 0 & 0 & 0 & 0 & 0 & 0 \\ 0 & 0 & 0 & 0 & 0 & 0 & 0 & 0 \\ 0 & 0 & 0 & 0 & 0 & 0 & 0 & 0 \\ 0 & 0 & 0 & 0 & 0 & 0 & 0 & 0 \\ 0 & 0 & 0 & 0 & k_{11}^4 & k_{12}^4 & k_{13}^4 & k_{14}^4 \\ 0 & 0 & 0 & 0 & k_{21}^4 & k_{22}^4 & k_{23}^4 & k_{24}^4 \\ 0 & 0 & 0 & 0 & k_{31}^4 & k_{32}^4 & k_{33}^4 & k_{34}^4 \\ 0 & 0 & 0 & 0 & k_{41}^4 & k_{42}^4 & k_{43}^4 & k_{44}^4 \end{bmatrix} \begin{Bmatrix} 0 \\ 0 \\ 0 \\ 0 \\ u_{3x} \\ u_{3y} \\ u_{4x} \\ u_{4y} \end{Bmatrix} \tag{10.71}
$$

Substitute back each of the element load vector expressions into expression (10.66), and the global stiffness matrix of the truss system is developed as

$$
\begin{Bmatrix} F_{1x} \\ F_{1y} \\ F_{2x} \\ F_{2y} \\ F_{3x} \\ F_{3y} \\ F_{4x} \\ F_{4y} \end{Bmatrix} =
$$

$$
\begin{bmatrix} k_{11}^1 & k_{12}^1 & k_{13}^1 & k_{14}^1 & 0 & 0 & 0 & 0 \\ k_{21}^1 & k_{22}^1 & k_{23}^1 & k_{24}^1 & 0 & 0 & 0 & 0 \\ k_{31}^1 & k_{32}^1 & k_{33}^1+k_{11}^2+k_{11}^3 & k_{34}^1+k_{12}^2+k_{12}^3 & k_{13}^3 & k_{14}^3 & k_{13}^2 & k_{14}^2 \\ k_{41}^1 & k_{42}^1 & k_{43}^1+k_{21}^2+k_{21}^3 & k_{44}^1+k_{22}^2+k_{22}^3 & k_{23}^3 & k_{24}^3 & k_{23}^2 & k_{24}^2 \\ 0 & 0 & k_{31}^3 & k_{32}^3 & k_{33}^3+k_{11}^4 & k_{34}^3+k_{12}^4 & k_{13}^4 & k_{14}^4 \\ 0 & 0 & k_{41}^3 & k_{42}^3 & k_{43}^3+k_{21}^4 & k_{44}^3+k_{22}^4 & k_{23}^4 & k_{24}^4 \\ 0 & 0 & k_{31}^2 & k_{32}^2 & k_{31}^4 & k_{32}^4 & k_{33}^2+k_{33}^4 & k_{34}^2+k_{34}^4 \\ 0 & 0 & k_{41}^2 & k_{42}^2 & k_{41}^4 & k_{42}^4 & k_{43}^2+k_{43}^4 & k_{44}^2+k_{44}^4 \end{bmatrix} \begin{Bmatrix} u_{1x} \\ u_{1y} \\ u_{2x} \\ u_{2y} \\ u_{3x} \\ u_{3y} \\ u_{4x} \\ u_{4y} \end{Bmatrix}
$$

$$\tag{10.72}$$

In general terms,

$$
\{F\}^{nodal} = [K]^{Global} \{U\}^{nodal} \tag{10.73}
$$

For this numerical problem, the known external loadings are $F_{2x} = 0$, $F_{2y} = 0$, $F_{4x} = 0$ and $F_{4y} = -450\ lbf$.

Also, the known boundary conditions are $u_{1x} = 0$, $u_{1y} = 0$, $u_{3x} = 0$ and $u_{3y} = 0$.

Substitute the stiffness k values from expressions (10.53) through (10.56) into the above expression (10.72) and solve for the unknowns (simultaneous equations). The results are as follows.

For the displacements:

$$u_{2x} = -1.61 \times 10^{-3}\ in. \tag{10.74}$$

$$u_{2y} = -6.6 \times 10^{-3}\ in. \tag{10.75}$$

$$u_{4x} = 0 \tag{10.76}$$

$$u_{4y} = -7.32 \times 10^{-3}\ in. \tag{10.77}$$

For the external loads:

$$F_{1x} = 674\ lbf \tag{10.78}$$

$$F_{1y} = 0 \tag{10.79}$$

$$F_{3x} = -674\ lbf \tag{10.80}$$

$$F_{3y} = 450\ lbf \tag{10.81}$$

The axial loads on each truss member are determined by using expressions (10.68), (10.69), (10.70) and (10.71) and application of the nodal displacements. The stress on each member is determined as follows.

For member (1):

$$f_{1x}^1 = k_{13}^1 u_{2x} + k_{14}^1 u_{2y} = (-418610)(-1.61 \times 10^{-3}) = 674\ \text{lbf} \tag{10.82}$$

$$\sigma^{(1)} = \frac{f_{1x}^1}{A^{(1)}} = \frac{674}{2.33} = 289\ \text{psi} \tag{10.83}$$

For member (2):

$$f_{2y}^2 = k_{21}^2 u_{2x} + k_{22}^2 u_{2y} + +k_{23}^2 u_{4x} + k_{24}^2 u_{4y}$$
$$= (626853)(-6.6 \times 10^{-3}) + (-626853)(-7.32 \times 10^{-3}) = 450\ lbf \tag{10.84}$$

$$\sigma^{(2)} = \frac{f_{2y}^2}{A^{(2)}} = \frac{450}{2.33} = 193\ \text{psi} \tag{10.85}$$

For member (3):

$$f_{2x}^3 = k_{11}^3 u_{2x} + k_{12}^3 u_{2y} + k_{13}^3 u_{3x} + k_{14}^3 u_{3y}$$
$$= (240546)(-1.61 \times 10^{-3}) + (-160667)(-6.6 \times 10^{-3}) = 674 \; lbf \tag{10.86}$$

$$f_{2y}^3 = k_{21}^3 u_{2x} + k_{22}^3 u_{2y} + k_{23}^3 u_{3x} + k_{24}^3 u_{3y}$$
$$= (-160667)(-1.61 \times 10^{-3}) + (107313)(-6.6 \times 10^{-3}) = -450 \; lbf \tag{10.87}$$

$$f^3 = \sqrt{(f_{2x}^3)^2 + (f_{2y}^3)^2} = 810 \; \text{lbf} \tag{10.88}$$

$$\sigma^{(3)} = \frac{f^3}{A^{(3)}} = \frac{810}{2.33} = 348 \; \text{psi} \tag{10.89}$$

For member (4):

$$f_{3x}^4 = k_{11}^4 u_{3x} + k_{12}^4 u_{3y} + k_{13}^4 u_{4x} + k_{14}^4 u_{4y} = 0 \tag{10.90}$$

$$\sigma^{(4)} = \frac{f_{3x}^4}{A^{(4)}} = 0 \tag{10.91}$$

10.3 STRESS ON BEAM MEMBERS

Consider a beam of the length L with the shear forces F_1 and F_2 and the moments M_1 and M_2 applied to its ends, as shown by Figure 10.7, below.

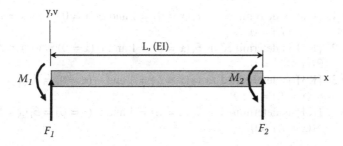

FIGURE 10.7 The beam member deformation.

The moment along the length of the beam at any diatance x is defined by elementary mechanics of solid as

$$M(x) = F_1 x - M_1 \tag{10.92}$$

The curvature equation for this beam with respect to the moment is

$$\frac{d^2v}{dx^2} = \frac{M(x)}{EI} = \frac{F_1x - M_1}{EI} \tag{10.93}$$

By integration, the slope of the beam is defined as

$$\frac{dv}{dx} = \frac{1}{E}\left(\frac{F_1x^2}{2} - M_1x\right) + C_1 = \theta \tag{10.94}$$

Likewise, by a second integration the deflection of the beam is defined as

$$v = \frac{1}{E}\left(\frac{F_1x^3}{6} + \frac{M_1x^2}{2}\right) + C_1x + C_2 \tag{10.95}$$

Now, knowing that the load-deflection relation for the beam is defined as

$$\begin{Bmatrix} F_1 \\ M_1 \\ F_2 \\ M_2 \end{Bmatrix} = [\{K1\}\{K2\}\{K3\}\{K4\}]\begin{Bmatrix} v_1 \\ \theta_1 \\ v_2 \\ \theta_2 \end{Bmatrix} \tag{10.96}$$

the elements of the beam stiffness matrix can be determined by application of the unity deflection boundaries, i.e.

Case 1. $\{K1\}$ is determined as $v_1(x = 0) = 1$ and $\theta_1(x = 0) = v_2(x = L) = \theta_2(x = L) = 0$.

Case 2. $\{K2\}$ is determined as $\theta_1(x = 0) = 1$ and $v_1(x = 0) = v_2(x = L) = \theta_2(x = L) = 0$.

Case 3. $\{K3\}$ is determined as $v_2(x = L) = 1$ and $v_1(x = 0) = \theta_1(x = 0) = \theta_2(x = L) = 0$.

Case 4. $\{K4\}$ is determined as $\theta_2(x = L) = 1$ and $v_1(x = 0) = \theta_1(x = 0) = v_2(x = L) = 0$.

For case 1, apply the respective boundary conditions into equations (10.94) and (10.95) and determine the constants of the integration:

$$v_1(x = 0) = 1, \quad 1 = \frac{1}{EI}\left(\frac{F_1(0)^3}{6} + \frac{M_1(0)^2}{2}\right) + C_1(0) + C_2 \rightarrow C_2 = 1 \tag{10.97}$$

$$\theta_1(x = 0) = 0, \ 0 = \frac{1}{EI}\left(\frac{F_1(0)^2}{2} - M_1(0)\right) + C_1 \rightarrow C_1 = 0 \quad (10.98)$$

$$v_2(x = L) = 0, \ 0 = \frac{1}{EI}\left(\frac{F_1(L)^3}{6} + \frac{M_1(L)^2}{2}\right) + 1 \rightarrow -EI = \frac{F_1 L^3}{6} - \frac{M_1 L^2}{2}$$
$$(10.99)$$

$$\theta_2(x = L) = 0, \ 0 = \frac{1}{EI}\left(\frac{F_1(L)^2}{2} - M_1(L)\right) \rightarrow M_1 = \frac{F_1 L}{2} \quad (10.100)$$

Substitute (10.100) into (10.99) and one would have

$$-EI = \frac{F_1 L^3}{6} - \frac{F_1 L^3}{4} \rightarrow F_1 = \frac{12EI}{L^3} \quad (10.101)$$

Substitute (10.101) into (10.100) and one would have

$$M_1 = \frac{F_1 L}{2} = \frac{6EI}{L^2} \quad (10.102)$$

By equilibrium,

$$F_1 + F_2 = 0 \rightarrow F_2 = -F_1 \quad (10.103)$$

Thus,

$$F_2 = -\frac{12EI}{L^3} \quad (10.104)$$

Likewise, by equilibrium

$$M_1 + M_2 + FL_1 = 0 \rightarrow M_2 = F_1 L - M_1 \quad (10.105)$$

Thus,

$$M_2 = \frac{12EI}{L^2} - \frac{6EI}{L^2} = \frac{6EI}{L^2} \quad (10.106)$$

Hence, by comparison of expressions (10.101), (10.102), (10.104) and (10.106) with equation (10.96), one would find that the first vector of the stiffness matrix is

$$\{K1\} = \begin{Bmatrix} \frac{12EI}{L^3} \\ \frac{6EI}{L^2} \\ \frac{-12EI}{L^3} \\ \frac{6EI}{L^2} \end{Bmatrix} \quad (10.107)$$

Similarly, for case 2, apply the respective boundary conditions into equations (10.94) and (10.95) and determine the constants of the integration. Then, determine the corresponding shear and moment expressions for unity slope. One would have

$$C_1 = 1 \tag{10.108}$$

$$C_2 = 0 \tag{10.109}$$

and

$$F_1 = \frac{6EI}{L^2} \tag{10.110}$$

$$M_1 = \frac{4EI}{L} \tag{10.111}$$

$$F_2 = -\frac{6EI}{L^2} \tag{10.112}$$

$$M_2 = \frac{2EI}{L} \tag{10.113}$$

Likewise, by comparison of expressions (10.110), (10.111), (10.112) and (10.113) with equation (10.96), one would find that the second vector of the stiffness matrix is

$$\{K2\} = \left\{ \begin{array}{c} \dfrac{6EI}{L^2} \\ \dfrac{4EI}{L} \\ \dfrac{-6EI}{L^2} \\ \dfrac{2EI}{L} \end{array} \right\} \tag{10.114}$$

Similarly, for case 3, apply the respective boundary conditions into equations (10.94) and (10.95) and determine the constants of the integration. Then, determine the corresponding shear and moment expressions for unity displacement. One would have

$$C_1 = 0 \tag{10.115}$$

$$C_2 = 0 \tag{10.116}$$

and

$$F_1 = -\frac{12EI}{L^3} \tag{10.117}$$

$$M_1 = -\frac{6EI}{L^2} \qquad (10.118)$$

$$F_2 = \frac{12EI}{L^3} \qquad (10.119)$$

$$M_2 = -\frac{6EI}{L^2} \qquad (10.120)$$

Likewise, by comparison of expressions (10.117), (10.118), (10.119) and (10.120) with equation (10.96), one would find that the third vector of the stiffness matrix is

$$\{K3\} = \begin{Bmatrix} -\frac{12EI}{L^3} \\ -\frac{6EI}{L^2} \\ \frac{12EI}{L^3} \\ -\frac{6EI}{L^2} \end{Bmatrix} \qquad (10.121)$$

Similarly, for case 4, apply the respective boundary conditions into equations (10.94) and (10.95) and determine the constants of the integration. Then, determine the corresponding shear and moment expressions for unity slope. One would have

$$C_1 = 0 \qquad (10.122)$$

$$C_2 = 0 \qquad (10.123)$$

and

$$F_1 = \frac{6EI}{L^2} \qquad (10.124)$$

$$M_1 = \frac{2EI}{L} \qquad (10.125)$$

$$F_2 = -\frac{6EI}{L^2} \qquad (10.126)$$

$$M_2 = \frac{4EI}{L} \qquad (10.127)$$

Likewise, by comparison of expressions (10.124), (10.125), (10.126) and (10.127) with equation (10.96), one would find that the fourth vector of the stiffness matrix is

$$\{K4\} = \begin{Bmatrix} \dfrac{6EI}{L^2} \\[2mm] \dfrac{2EI}{L} \\[2mm] -\dfrac{6EI}{L^2} \\[2mm] \dfrac{4EI}{L} \end{Bmatrix} \tag{10.128}$$

The element stiffness matrix can be assembled together as

$$[K] = \begin{bmatrix} \dfrac{12EI}{L^3} & \dfrac{6EI}{L^2} & \dfrac{-12EI}{L^3} & \dfrac{6EI}{L^2} \\[2mm] \dfrac{6EI}{L^2} & \dfrac{4EI}{L} & \dfrac{-6EI}{L^2} & \dfrac{2EI}{L} \\[2mm] \dfrac{-12EI}{L^3} & \dfrac{-6EI}{L^2} & \dfrac{12EI}{L^3} & \dfrac{-6EI}{L^2} \\[2mm] \dfrac{6EI}{L^2} & \dfrac{2EI}{L} & \dfrac{-6EI}{L^2} & \dfrac{4EI}{L} \end{bmatrix} \tag{10.129}$$

This matrix is only valid for vertical translation and in-plane rotation. Now, consider a beam under torsion-loading, as shown by Figure 10.8, below.

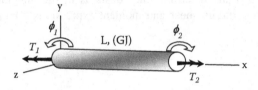

FIGURE 10.8 The torsional deformation of beam.

Under the applied loading, the beam twists a total amount of ϕ ($\phi = \phi_1 - \phi_2$). The strain on the beam is defined as

$$\gamma = \frac{r}{L}\phi \tag{10.130}$$

and the shear stress is defined in terms of the total torque on beam as

$$\tau = \frac{Tr}{J} \tag{10.131}$$

By Hooke's law, the stress and strain are related as such:

$$\tau = G\gamma \tag{10.132}$$

Thus, by substitution of (10.130) and (10.131) into Hooke's relation, one would have

$$\frac{Tr}{J} = G\frac{r}{L}\phi \tag{10.133}$$

or for twist

$$\phi = \frac{TL}{JG} \tag{10.134}$$

Solving for torque as a function of twist, one would have

$$T = \frac{JG}{L}\phi \tag{10.135}$$

where $\frac{JG}{L}$ is the stiffness of the beam. Let k_T term represent this stiffness:

$$k_T = \frac{JG}{L} \tag{10.136}$$

Then, expression (10.135) can be rewritten as

$$T = k_T\phi \tag{10.137}$$

Now, consider the beam under equilibrium conditions. The summation of torques on the beam must equal zero.

$$\sum T = 0, \ T_1 + T_2 = 0 \text{ or } T_1 = -T_2 \tag{10.138}$$

The total beam twist can be represented by the individual end rotations:

$$\phi = \phi_1 - \phi_2 \tag{10.139}$$

Apply expression (10.137) and the torque on the beam is

$$T_2 = -T_1 = k_T(\phi_2 - \phi_1) \tag{10.140}$$

or

$$T_1 = \phi_1 k_T - \phi_2 k_T \tag{10.141}$$

and

$$T_2 = \phi_2 k_T - \phi_1 k_T \tag{10.142}$$

Expressions (10.141) and (10.142) can be represented in the matrix form as

$$\begin{Bmatrix} T_1 \\ T_2 \end{Bmatrix} = \begin{bmatrix} k_T & -k_T \\ -k_T & k_T \end{bmatrix} \begin{Bmatrix} \phi_1 \\ \phi_2 \end{Bmatrix} \tag{10.143}$$

which defines the twist behavior of the beam under torsion.

One can combine the stiffness matrix (10.129) and (10.143) to represent the total beam behavior under shear, torsion and moment all together as follows:

$$\begin{Bmatrix} F_1 \\ T_1 \\ M_1 \\ F_2 \\ T_2 \\ M_2 \end{Bmatrix} = \begin{bmatrix} \frac{12EI}{L^3} & 0 & \frac{6EI}{L^2} & \frac{-12EI}{L^3} & 0 & \frac{6EI}{L^2} \\ & \frac{JG}{L} & 0 & 0 & \frac{-JG}{L} & 0 \\ & & \frac{4EI}{L} & \frac{-6EI}{L^2} & 0 & \frac{2EI}{L} \\ & & & \frac{12EI}{L^3} & 0 & \frac{-6EI}{L^2} \\ & & & & \frac{JG}{L} & 0 \\ \textit{Symmetric} & & & & & \frac{4EI}{L} \end{bmatrix} \begin{Bmatrix} v_1 \\ \phi_1 \\ \theta_1 \\ v_2 \\ \phi_2 \\ \theta_2 \end{Bmatrix} \tag{10.144}$$

This relation is to be used the same way as the relation (10.43) for truss members was used. It is used to build up the beam system of simultaneous equations, which can be solved to determine the beam displacements and rotations.

10.4 ACCURATE FINITE ELEMENT ANALYSIS OF PLATES

One can derive an expression representing the stress field in a plate shown in Figure 10.9 from *Airy* stress function. The stress expression derived would be in polar coordinate where σ_r (equation 10.145) represents the stress in radial direction to the center hole. Likewise, σ_θ (equation 10.146) represents the stress in the angular direction at the circumference of the center hole. $\tau_{r\theta}$ (equation 10.147) would represent the shearing stress in polar coordinates. In all three expressions, r represents the radial location, a represents the center hole radius, σ represents the normal stress due to normal loading and θ represents the angular location from horizontal, as shown by Figure 10.9.

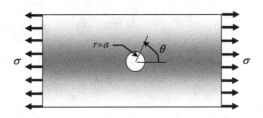

FIGURE 10.9 Plate model with center hole.

$$\sigma_r = \frac{\sigma}{2}\left[\left(1 - \frac{a^2}{r^2}\right) + \left(1 + \frac{3a^4}{r^4} - \frac{4a^2}{r^2}\right)\cos 2\theta\right] \qquad (10.145)$$

$$\sigma_\theta = \frac{\sigma}{2}\left[\left(1 + \frac{a^2}{r^2}\right) - \left(1 + \frac{3a^4}{r^4}\right)\cos 2\theta\right] \qquad (10.146)$$

$$\tau_{r\theta} = -\frac{\sigma}{2}\left(1 - \frac{3a^4}{r^4} + \frac{2a^2}{r^2}\right)\sin 2\theta \qquad (10.147)$$

Further, the stress in polar coordinates can be expressed in terms of the stresses in the Cartesian coordinate system, as shown by equations (10.148), (10.149) and (10.150). σ_x represents the normal stress, σ_y represents the transverse stress and τ_{xy} represents the shear stress in the Cartesian coordinate system.

$$\sigma_r = \sigma_x \cos^2\theta + \sigma_y \sin^2\theta + 2\tau_{xy}\sin\theta\cos\theta \qquad (10.148)$$

$$\sigma_\theta = \sigma_x \sin^2\theta + \sigma_y \cos^2\theta - 2\tau_{xy}\sin\theta\cos\theta \qquad (10.149)$$

$$\tau_{r\theta} = (\sigma_y - \sigma_x)\sin\theta\cos\theta + \tau_{xy}(\cos^2\theta - \sin^2\theta) \qquad (10.150)$$

The expressions in equations (10.148), (10.149) and (10.150) can be formulated in matrix and vector forms and the simultaneous equations can be solved for the stresses in the Cartesian coordinate. From the Cartesian stress expressions, the principal stress values can be determined by expression in equation (10.151).

$$\sigma_{1,2} = \frac{\sigma_x + \sigma_y}{2} \pm \sqrt{\left(\frac{\sigma_x - \sigma_y}{2}\right)^2 + \tau_{xy}^2} \qquad (10.151)$$

Finally, one can derive the Von-Mises stress (σ_{von}) representing the stress field on the plate by expressions in equation (10.152) where, σ_1, σ_2 and σ_3 are the principal stresses determined by equation (10.151). This Von-Mises stress would be used to draw the theoretical closed-form stress solution of the plate with a central hole.

$$\sigma_{von-mises} = \sqrt{\frac{(\sigma_1 - \sigma_2)^2 + (\sigma_2 - \sigma_3)^2 + (\sigma_1 - \sigma_3)^2}{2}} \qquad (10.152)$$

A quarter symmetric model of a thin plate (0.2 inch) with center hole (4 in. diameter), is built with 4-node shell elements in ANSYS. The plate full dimension is 20 in. by 20 in. Four distinct meshes are generated to elaborate the

(a)   (b)

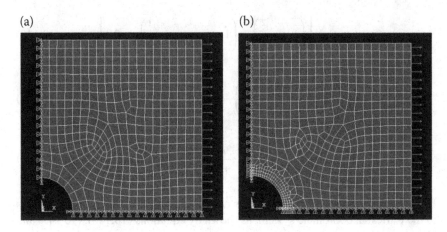

FIGURE 10.10   a) free-mesh b) high-density free-mesh using membrane shell elements.

shape dependency of the FEA results. The first model is built with free-mesh of shell elements that are of fairly equal sides (LD Free-mesh). Figure 10.10(a) illustrates this unstructured mesh. The same model is built from a high-density free-mesh (HD Map-mesh). In this mesh, in the immediate proximity of the hole area, where there is stress concentration, the mesh density is higher. Figure 10.10(b) illustrates this final mesh. A third model is built with mapped-mesh of shell elements that are of almost equal sizes but have high aspect ratios (HAR Map-mesh). That is the ratio of the long side of the element to the shorter side if it is large (slender shape). Also, these elements follow a uniform pattern. Figure 10.11(a) illustrates the shape and pattern of this structured mesh. The fourth model is built with mapped-mesh of shell elements that have a low aspect ratio. That is, the ratio of the longer side of the elements to the shorter side is

(a)   (b)

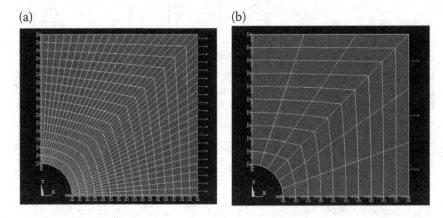

FIGURE 10.11   a) high aspect ratio mapped-mesh b) low aspect ratio mapped-mesh using membrane shell elements.

small (LAR Map-mesh). Figure 10.11(b) clearly illustrates the shape and pattern of this structured mesh. These four mesh models adequately summarize all of the possible mesh patterns that could be generated to mesh this plate geometry.

Symmetric properties are used in order to model only one fourth of the full plate model. Symmetric boundary conditions are used at the left and bottom sides of the plate models. For applied loading, a 50 lb/in. uniform line pressure is applied to the right side of the model (depicted by red arrows in the meshed figure). This 50 lb/in. line loading is equivalent to a quasi-static 1000 lbf tensile load applied to the plate thickness, in the right side. The top side of the plate model is free from restraints.

Table 10.2 below is used to index the four different mesh type models with respect to the number of elements, element size and aspect ratio.

TABLE 10.2
The Mesh Type Index for All Models

Mesh Type	HD Mesh	LD Free Mesh	LAR Map Mesh	HAR Map Mesh
Number of plate elements	482	395	60	640
Average Size(in^2)	0.08	0.25	3	0.5
% plate elements with high aspect ratio	0	0	40	100
% plate elements with low aspect ratio	100	100	60	0

The "HD Mesh" is the high-density free mesh, "LD Free Mesh" is the low-density free mesh, "LAR Map Mesh" is the low aspect ratio mapped mesh and "HAR Map Mesh" is the high aspect ratio mapped mesh. This comparison table materializes the mesh represented in Figures 10.10 and 10.11 preceding.

Figures 10.12 through 10.15 represent the whole-field displacements for the free-mesh, high aspect ratio mapped-mesh, low aspect ratio mapped-mesh and the high-density mesh, respectively. All four models produce identical displacement counters of the plate model with the same maximum displacements. The displacement magnitudes indicate that the model is stiff, and contour patterns indicate all four models are strained in the same manner.

The stress contours representing the stress field of the plate are plotted in Figures 10.16 through 10.19. All of the plots, from the four different possible meshes, indicate a maximum Von-Mises magnitude at the 90° location around the hole at r = 2 in. This is a stress concentration location, which is also expected by theoretical values where displacement in the x-direction is considered to be constrained. Comparing all models the stress contours over the entire plate model are similar in pattern. The only variations are the mangitudes of the stress levels developed in the plate stress field. The model with a denser mesh captures the stress levels more accurately, as it will be shown by Figure 10.20 later. The

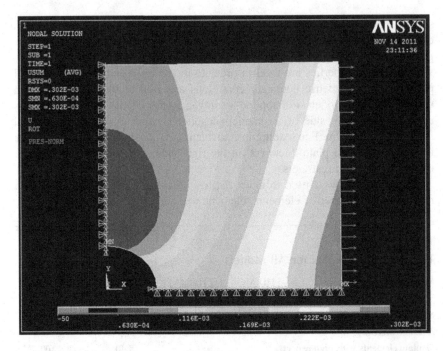

FIGURE 10.12 Displacement contour for the unstructured free-mesh model.

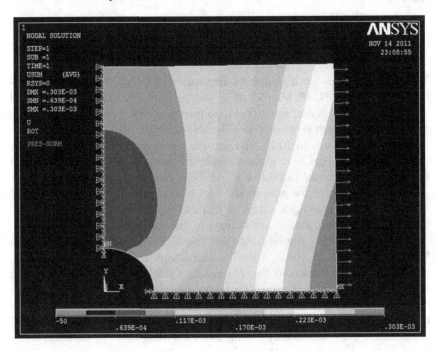

FIGURE 10.13 Displacement contour for the structured high aspect ratio mapped-mesh model.

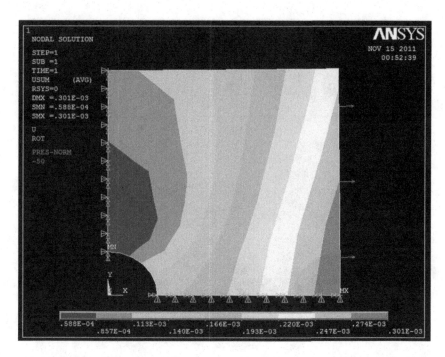

FIGURE 10.14 Displacement contour for the structured low aspect ratio mapped-mesh model.

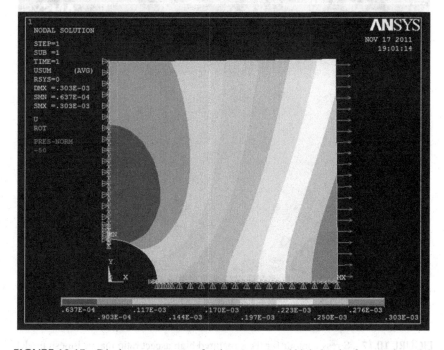

FIGURE 10.15 Displacement contour for the unstructured high density free-mesh model.

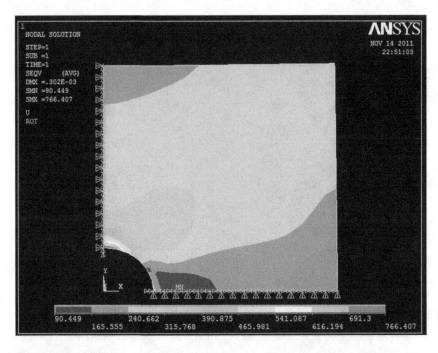

FIGURE 10.16   Stress contour for the unstructured free-mesh model.

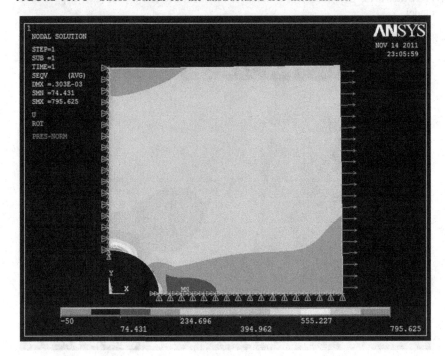

FIGURE 10.17   Stress contour for the structured high aspect ratio mapped-mesh model.

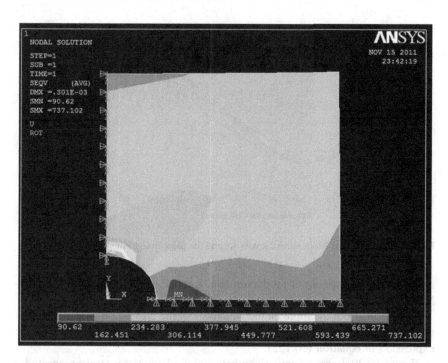

FIGURE 10.18 Stress contour for the structured low aspect ratio mapped-mesh model.

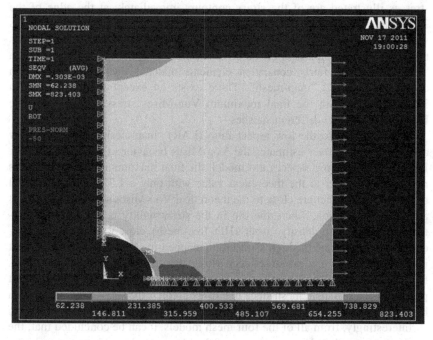

FIGURE 10.19 Stress contour for the unstructured high density free-mesh model.

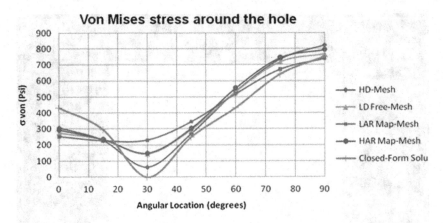

FIGURE 10.20   Von-Mises stress levels around the plate hole radius for different meshs.

models with mapped-mesh produce a less accurate representation of the Von-Mises stresses.

The results from the finite element analysis of the previous four mesh models are compared with the theoretical Von-Mises stress levels derived by using the expression in equation (10.152).

Figure 10.20 draws a comparison between the Von-Mises stress-magnitude-results developed based on the different mesh generations. The Von-Mises stresses illustrated are of the stress contours immediately at the edge of the 2 in. radius hole section (r = 2 in.). The figure compares different mesh results with one another and to the results developed from closed-form stress analysis theory. A close comparison of the different mesh models used indicates that a "free-mesh" with fairly consistent elements matches the results of a high aspect ratio (HAR) "map-mesh". There exists an exception that there is a 3.65% difference in the final maximum Von-Mises stress due to the same loading for the two different meshes.

In comparison to the low aspect ratio (LAR) "map-mesh" results, the low aspect ratio mesh over estimates the Von-Mises behavior of the plate slightly. However, for the low aspect ratio model, the final maximum Von-Mises stress magnitude is closer to the theoretical value with only a 1.72% difference. All three meshes together are close to the theoretical Von-Mises stress values except at 30° angle location, where the dip in the stress nullity is not very well determined. The high-density mesh (HD free-mesh) closely follows the low-density free-mesh and the high aspect ratio mapped-mesh with the exception that it estimates the theoretical stress deep closer than the other three models. This mesh overestimates the theoretical final maximum stress level by 9.73%. Yet, it proves to be a sufficient mesh since in closely predicts the stress behavior as seen by Figure 10.20.

Interestingly, from all of the four mesh models, it can be concluded that, the proper aspect ratio-ed mapped-mesh of the elements alone, does not guaranty accurate results. An adequate mesh density is required to predict the stress behavior of a structure. Also, it could be concluded that low aspect ratio meshes are a good tool for determining margins of safety, since they may closely predict the maximum stress levels.

Most finite element analysis packages in some way provide meshing tools that generate well behaved (moderate aspect ratio) elements. Yet, none of them provide for an automatic mesh density measure. The analyst has to select and control the mesh density of the region of the interest being analyzed. This has to be done either by preliminary FEA trial runs detecting the stress concentration regions or pre-determination of the specified region by theoretical means. In either way, also the degree mesh density refinement has to be dealt with that only comes with FEA experience of similar structures.

To elaborate on the far-field accuracy of the most accurate mesh (the high-density mesh-HD mesh), the plot of the normal stress vs. the vertical distance from the hole to the horizontal edge of the plate in y-direction (at the left symmetry line) is plotted in Figure 10.21 below. The FEA results from the high-density mesh minics the theoretical values of the normal stress on the plate accurately. It is conclusive that the high-desnity mesh also is a good far-field indicator.

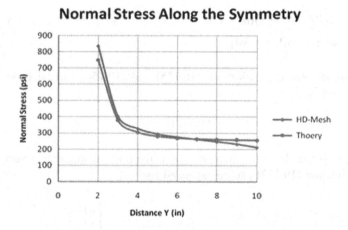

Normal Stress Along the Symmetry

FIGURE 10.21 The normal stress vs. the vertical distance to the top.

To explain the theoretical reasoning for having less accuracy for high aspect ratio elements, it is essential to understand the computational method for the finite element modeling. One can begin by understanding the isoparametric formulation of elements [7]. For every element as illustrated by Figure 10.22, there is a linear displacement function defined in horizontal direction u and vertical direction v.

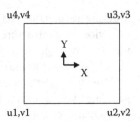

FIGURE 10.22 Four-node rectangular element.

$$u = \beta 1 + \beta 2x - \beta 3y + \beta 4xy \tag{10.153}$$

$$v = \beta 5 + \beta 6x - \beta 7y + \beta 8xy \tag{10.154}$$

Using the boundary conditions at each node, the β's can be eleminated and displacement functions can be expressed as:

$$u = (1/4bh)[(b - x)(h - y)u1 + (b + x)(h - y)u2 + (b + x)(h + y)u3$$
$$+ (b - x)(h + y)u4] \tag{10.155}$$

$$v = (1/4bh)[(b - x)(h - y)v1 + (b + x)(h - y)v2 + (b + x)(h + y)v3$$
$$+ (b - x)(h + y)v4] \tag{10.156}$$

The displacement expressions in (10.155) and (10.156) can be expressed in matrix form as:

$$\{U\} = [N]\{d\} \tag{10.157}$$

where [N] is the shape function and {d} is the nodal displacement vector.
The relation (10.157) can be expanded and

$$
\begin{Bmatrix} u \\ v \end{Bmatrix} = \begin{bmatrix} N1\ 0\ N2\ 0\ N3\ 0\ N4\ 0 \\ 0\ N1\ 0\ N2\ 0\ N3\ 0\ N4 \end{bmatrix} \begin{Bmatrix} u1 \\ v1 \\ u2 \\ v2 \\ u3 \\ v3 \\ u4 \\ v4 \end{Bmatrix} \tag{10.158}
$$

where Ni coefficients are the shape functions in terms of x, y, b and h.

Now, the strain in terms of the displacement is:

$$\begin{Bmatrix} \varepsilon x \\ \varepsilon y \\ \gamma xy \end{Bmatrix} = \begin{Bmatrix} \frac{\partial u}{\partial x} \\ \frac{\partial v}{\partial y} \\ \frac{\partial u}{\partial y} + \frac{\partial v}{\partial x} \end{Bmatrix} \quad (10.159)$$

which is

$$\begin{Bmatrix} \varepsilon x \\ \varepsilon y \\ \gamma xy \end{Bmatrix} =$$

$$\begin{bmatrix} -(h-y) & 0 & (h-y) & 0 & (h+y) & 0 & -(h+y) & 0 \\ 0 & -(b-x) & 0 & -(b+x) & 0 & (b+x) & 0 & (b-x) \\ -(b-x) & -(h-y) & -(b+x) & (h-y) & (b+x) & (h+y) & (b-x) & -(h+y) \end{bmatrix} \begin{Bmatrix} u1 \\ v1 \\ u2 \\ v2 \\ u3 \\ v3 \\ u4 \\ v4 \end{Bmatrix}$$

$$(10.160)$$

where for a plane stress condition, the stress vector $\{\sigma\}$ in terms of the element strains can be derived as:

$$\{\sigma\} = [D]\{\varepsilon\} \quad (10.161)$$

where D is the material stifness matrix:

$$D := \frac{E}{1-v^2} \cdot \begin{bmatrix} 1-v & v & 0 \\ v & 1 & 0 \\ 0 & 0 & \frac{(1-v)}{2} \end{bmatrix} \quad (10.162)$$

Now, from expression (10.161) a comparison between the stress results from a high aspect ratio element and a low aspect ratio element can be computed, and one would have a higher stress value for the high aspect ratio element.

10.5 FINITE ELEMENT ANALYSIS RESULTS CORRELATIONS

It is important that in the FEA modeling representation of a design, correlation of the results to practical experimental or analytical solutions are carried out. Oftentimes, in FEA modeling due to discrepancies between the boundary conditions selected for the FEA model versus the actual design constraints, the results may not be accurate and reliable. To check for such discrepancies, one could start the model as a simple basic model representation and apply loads and boundary conditions to check for results and comparison with either experimental or closed form analytical solution results. Once a simplified model of the design is proven to be viable and checks out to hold the proper boundary conditions that provide an equilibrium load model that balances the load inputs to the load outputs, then the FEA model can be detailed and expanded. Mesh density studies and proper element shape sizing, discussed in previous section 10.4, has to come secondary after the boundary conditions are validated.

Proper use of the element numbers that capture the expected outputs have to be carefully considered in FEA modeling. As an example, if one is modeling a thin plates substructure and bending properties of thin plate through the thickness need to be captured in order to have accurate bending stress results, through the thickness element numbers are an essential part of the FEA modeling. In this case, at least three elements through the thickness are recommended to be practical.

Transitioning of 2-D elements to 3-D elements need to be carefully evaluated, and normally it is not recommended in an FEA model. In the 2-D shell element type models, you may have all 6 degress of freedom (DOF) for an output, as in the 3-D solid elements you may have only 3 DOF for the output (3 translational DOF versus all 6 DOF, which includes the translational and rotational DOFs).

Likewise, when using bar or beam elements, stress analyst has to consider if bending stresses are of interest for the output or not. Beam elements generate results capturing the bending stresses, whereas rod elements may not allow such results. If connection pins are to be modeled and are connected using the RBE2 and RBE3 elements that transfer loads from plate and/or solid elements, then beam elements should be used for pins.

10.6 DETERMINATION OF FASTENER STIFFNESS FOR FEA

Modeling bolt fasteners in finite element analysis is an essential and important part of modeling since the fasteners are usually the gatekeepers for transferring loads to the main structure under stress analysis investigation and often represent the interface attachment points for a substructure to and from a primary structure. For that reason, when one uses the CBUSH type elements to represent the bolt fasteners in an FEM model, the stiffness of the bolt is required as an input.

Hereby, the determination of fastener stiffness analytically is explored. Assume that the k variable represents stiffness, δ represents the change in length (L) of the bolt under an applied loading of P. One knows that:

$$P = k(\delta) \tag{10.163}$$

Thus,

$$k = \frac{P}{\delta} \tag{10.164}$$

Also,

$$\sigma = E(\varepsilon) \tag{10.165}$$

Thus,

$$\varepsilon = \frac{\sigma}{E} \tag{10.166}$$

Knowing that,

$$\sigma = \frac{P}{A} \tag{10.167}$$

$$\varepsilon = \frac{\sigma}{E} \varepsilon = \frac{\sigma}{E}$$

Then,

$$\varepsilon = \frac{P}{AE} \tag{10.168}$$

Also knowing that,

$$\varepsilon = \frac{\delta}{L} \tag{10.169}$$

By equity, equations (10.168) and (10.169) would have:

$$\delta = \frac{PL}{AE} \tag{10.170}$$

Substituting back into the stiffness equation (10.164), then,

$$k = \frac{AE}{L} \tag{10.171}$$

Now, for a bolt that has a threaded section and an unthreaded section, give the subscripts T for threaded section and UT for the unthreaded section, and one would have:

$$k_T = \frac{A_T \cdot E}{L_T} \tag{10.172}$$

and

$$k_{UT} = \frac{A_{UT} \cdot E}{L_{UT}} \tag{10.173}$$

For a bolt, the threaded section and the unthreaded section are along the same direction on top of each other; thus, the equivalent stiffness k_{bolt} is:

$$\frac{1}{k_{bolt}} = \frac{1}{k_T} + \frac{1}{k_{UT}} \tag{10.174}$$

or

$$k_{bolt} = \frac{k_T \cdot k_{UT}}{L_T + k_{UT}} \qquad (10.175)$$

Use this equation (10.175) in conjunction with equations (10.171) and (10.173) to calculate the stiffness of the bolt for the CBUSH model input in the FEA.

Problems

1. For the truss system shown, determine the joint displacements and the member stresses assuming all members are made from square 0.5 in. by 0.5 in. bars. The bars are made out of steel material ($E = 29 \times 10^6$ psi).

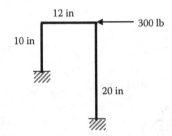

2. For the truss system shown, determine the joint displacements and the member stresses assuming all members are made from round 1 in. diameter bars. The bars are made out of steel material ($E = 29 \times 10^6$ psi).

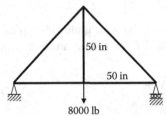

3. For the truss system shown, determine the joint displacements and the member stresses assuming all members are made from square 1 in. by 1 in. bars. The bars are made out of steel material ($E = 29 \times 10^6$ psi).

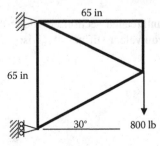

4. For the truss system shown, determine the joint displacements and the member stresses assuming all members are made from square 1.5 in. by 1.5 in. bars. The bars are made out of steel material ($E = 29 \times 10^6$ psi).

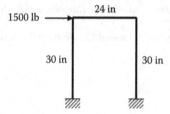

5. For the beam system shown, determine the stiffness matrix for vertical translation and in-plane rotation. Also, determine the forces in each element. The beams are made out of steel material ($E = 29 \times 10^6$ psi).

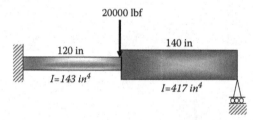

6. For the beam system shown, determine the stiffness matrix for vertical translation and in-plane rotation. Determine the tip displacement. Also, determine the forces in each element. The beams are made out of steel material ($E = 29 \times 10^6$ psi).

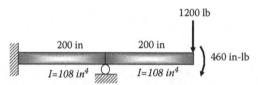

7. Using the expressions shown in equations (10.148), (10.149) and (10.150), solve for the state of stresses in the Cartesian coordinate system.

REFERENCES

Asmar, G., Chakar, E., "Analysis of an isotropic plate containing three identical circular holes arranged in a triangular configuration", Advances in Computational Tools for Engineering Applications Conference, 2009.

El-Sawy, K.M., Martini, M.I., "Stability of biaxially loaded square plates with single central holes", *Ships and Offshore Structures Conf.*, 2010.

Logan, D.L., *Finite Element Methods*, 2002. California: Wadsworth.

Madenci, E., Guven, I., *The Finite Element Method and Applications in Engineering Using ANSYS*, 2006. Boston, MA: Springer.

Phase 2; "Mesh generation tools", http://www.rockscience.com, 2011.

Picasso, M., "Adaptive finite elements with large aspect ratio based on an anisotropic error estimator involving first order derivatives", *Journal of Computational Mechnaics*, 2005. Elsevier.

Ugural, A.C., Fenster, S.K., *Advanced Strength and Applied Elasticity*, 1995. New Jersey: Prentice Hall.

11 Composite Analysis Theory

11.1 INTRODUCTION

The application of composite materials for design of various structural systems is becoming more common everyday. The composites are generally used because they can provide improvement of the mechanical properties such as strength, stiffness and toughness. Thus, stress analysis of composite materials will be required for design with such material. For that purpose, this chapter first provides the fundamental calculation theory necessary to evaluate composite lamina (single layer of laminated composites). Once the lamina behavior is determined, then, the stress analysis methods for composite laminate structures (bonded layers of lamina) are presented here.

11.2 ORTHOTROPIC LAMINA

The composite lamina by definition is an orthotropic material that has properties that are different in each material direction. The material direction parallel to the fiber direction is normally referred to as the longitudinal direction, and the material direction perpendicular to the fiber direction is normally referred to as the transverse direction. Figure 11.1 illustrates this. It is accurate to indicate that, the properties of the lamina in the longitudinal direction are governed by the fiber properties and the properties of the lamina in the transverse direction are governed by the matrix (epoxy) properties.

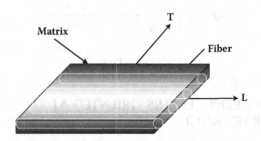

FIGURE 11.1 The composite lamina schematic.

DOI: 10.1201/9781003311218-11

Having said that and using the annotation of L for longitudinal direction and T for the transverse direction, one can denote the strain-stress relationship for a lamina as follows:

$$\varepsilon_L = \frac{\sigma_L}{E_L} - \upsilon_{TL}\frac{\sigma_T}{E_T} \tag{11.1}$$

$$\varepsilon_T = \frac{\sigma_T}{E_T} - \upsilon_{LT}\frac{\sigma_L}{E_L} \tag{11.2}$$

$$\gamma_{LT} = \frac{\tau_{LT}}{G_{LT}} \tag{11.3}$$

where E, υ and G with their subscripts represent the material properties of the lamina in their respective direction. (Note: $\upsilon_{LT}E_T = \upsilon_{TL}E_L$)

Likewise, the lamina stresses can be derived from the above relations as

$$\begin{Bmatrix} \sigma_L \\ \sigma_T \\ \tau_{LT} \end{Bmatrix} = \begin{bmatrix} Q11 & Q12 & 0 \\ Q12 & Q22 & 0 \\ 0 & 0 & Q66 \end{bmatrix} \begin{Bmatrix} \varepsilon_L \\ \varepsilon_T \\ \gamma_{LT} \end{Bmatrix} \tag{11.4}$$

where, in matrix form, the matrix containing the Q elements is referred to as the stiffness matrix, and they are as follows:

$$Q11 = \frac{E_L}{1 - \upsilon_{LT}\upsilon_{TL}} \tag{11.5}$$

$$Q22 = \frac{E_T}{1 - \upsilon_{LT}\upsilon_{TL}} \tag{11.6}$$

$$Q12 = \frac{\upsilon_{LT}E_T}{1 - \upsilon_{LT}\upsilon_{TL}} = \frac{\upsilon_{TL}E_L}{1 - \upsilon_{LT}\upsilon_{TL}} \tag{11.7}$$

$$Q66 = G_{LT} \tag{11.8}$$

11.3 ORTHOTROPIC LAYERS ORIENTED AT AN ARBITRARY ANGLE

The stresses and strains of the lamina in any direction, as shown by Figure 11.2, in terms of the longitudinal and transverse stresses and strains are represented by the following transformation (plane stress):

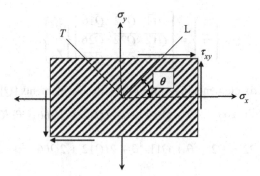

FIGURE 11.2 The lamina stresses at arbitrary angle.

$$\begin{Bmatrix} \sigma_x \\ \sigma_y \\ \tau_{xy} \end{Bmatrix} = \begin{bmatrix} c^2\theta & s^2\theta & -2s\theta c\theta \\ s^2\theta & c^2\theta & 2s\theta c\theta \\ s\theta c\theta & -s\theta c\theta & c^2\theta - s^2\theta \end{bmatrix} \begin{Bmatrix} \sigma_L \\ \sigma_T \\ \tau_{LT} \end{Bmatrix} \tag{11.9}$$

where c denotes cosine of an angle and s denotes sine of an angle.

Now, by substitution of expression (11.4) into this equation matrix, one would have

$$\begin{Bmatrix} \sigma_x \\ \sigma_y \\ \tau_{xy} \end{Bmatrix} = \begin{bmatrix} c^2\theta & s^2\theta & -2s\theta c\theta \\ s^2\theta & c^2\theta & 2s\theta c\theta \\ s\theta c\theta & -s\theta c\theta & c^2\theta - s^2\theta \end{bmatrix} \begin{bmatrix} Q11 & Q12 & 0 \\ Q12 & Q22 & 0 \\ 0 & 0 & Q66 \end{bmatrix} \begin{Bmatrix} \varepsilon_L \\ \varepsilon_T \\ \gamma_{LT} \end{Bmatrix} \tag{11.10}$$

Knowing that the strain transformation relation is

$$\begin{Bmatrix} \varepsilon_L \\ \varepsilon_T \\ \frac{\gamma_{LT}}{2} \end{Bmatrix} = \begin{bmatrix} c^2\theta & s^2\theta & 2s\theta c\theta \\ s^2\theta & c^2\theta & -2s\theta c\theta \\ -s\theta c\theta & s\theta c\theta & c^2\theta - s^2\theta \end{bmatrix} \begin{Bmatrix} \varepsilon_x \\ \varepsilon_y \\ \frac{\gamma_{xy}}{2} \end{Bmatrix} \tag{11.11}$$

Substitute this strain transformation (11.11) into expression (11.10) for the stress-strain relationship. As the result, one would have the following expression representing the state of stress in terms of strains in an arbitrary fiber angle:

$$\begin{Bmatrix} \sigma_x \\ \sigma_y \\ \tau_{xy} \end{Bmatrix} = \begin{bmatrix} c^2\theta & s^2\theta & -2s\theta c\theta \\ s^2\theta & c^2\theta & 2s\theta c\theta \\ s\theta c\theta & -s\theta c\theta & c^2\theta - s^2\theta \end{bmatrix} \begin{bmatrix} Q11 & Q12 & 0 \\ Q12 & Q22 & 0 \\ 0 & 0 & 2Q66 \end{bmatrix}$$
$$\begin{bmatrix} c^2\theta & s^2\theta & 2s\theta c\theta \\ s^2\theta & c^2\theta & -2s\theta c\theta \\ -s\theta c\theta & s\theta c\theta & c^2\theta - s^2\theta \end{bmatrix} \begin{Bmatrix} \varepsilon_x \\ \varepsilon_y \\ \frac{\gamma_{xy}}{2} \end{Bmatrix} \tag{11.12}$$

This relation can be represented in a simpler matrix form as:

$$\begin{Bmatrix} \sigma_x \\ \sigma_y \\ \tau_{xy} \end{Bmatrix} = \begin{bmatrix} \overline{Q11} & \overline{Q12} & \overline{Q16} \\ \overline{Q12} & \overline{Q22} & \overline{Q26} \\ \overline{Q16} & \overline{Q26} & \overline{Q66} \end{bmatrix} \begin{Bmatrix} \varepsilon_x \\ \varepsilon_y \\ \gamma_{xy} \end{Bmatrix} \qquad (11.13)$$

where these matrix elements in terms of the stiffness elements $[Q]$ are as follows:

$$\overline{Q11} = Q11c^4\theta + Q22s^4\theta + 2(Q12 + 2Q66)s^2\theta c^2\theta \qquad (11.14)$$

$$\overline{Q22} = Q22c^4\theta + Q11s^4\theta + 2(Q12 + 2Q66)s^2\theta c^2\theta \qquad (11.15)$$

$$\overline{Q12} = (Q11 + Q22 - 4Q66)s^2\theta c^2\theta + Q12(s^4\theta + c^4\theta) \qquad (11.16)$$

$$\overline{Q66} = (Q11 + Q22 - 2Q12 - 2Q66)s^2\theta c^2\theta + Q66(s^4\theta + c^4\theta) \quad (11.17)$$

$$\overline{Q16} = (Q11 - Q12 - 2Q66)c^3\theta s\theta - (Q22 - Q12 - 2Q66)c\theta s^3\theta) \qquad (11.18)$$

$$\overline{Q26} = (Q11 - Q12 - 2Q66)c\theta s^3\theta - (Q22 - Q12 - 2Q66)c^3\theta s\theta) \qquad (11.19)$$

It should be noted that the stresses along the lamina directions (L,T) can be presented in terms of the arbitrary axes stresses by inversing the relation shown by expression (11.9) as follows:

$$\begin{Bmatrix} \sigma_L \\ \sigma_T \\ \tau_{LT} \end{Bmatrix} = \begin{bmatrix} c^2\theta & s^2\theta & 2s\theta c\theta \\ s^2\theta & c^2\theta & -2s\theta c\theta \\ -s\theta c\theta & s\theta c\theta & c^2\theta - s^2\theta \end{bmatrix} \begin{Bmatrix} \sigma_x \\ \sigma_y \\ \tau_{xy} \end{Bmatrix} \qquad (11.20)$$

Example 11.1: For a lamina shown in the following figure, the strains are $\varepsilon_x = -400\ \mu\varepsilon$, $\varepsilon_y = 700\ \mu\varepsilon$ and $\gamma_{xy} = -450\ \mu\varepsilon$. Determine the normal stress components in the x and y direction and the shear stress. Also, determine the stresses and strains along the fiber and transverse matrix directions. ($E_L = 20 \times 10^6$ psi, $E_T = 1.3 \times 10^6$ psi, $G_{LT} = 1.03 \times 10^6$ psi, $v_{LT} = 0.3$)

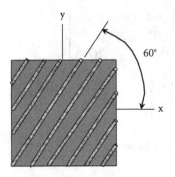

To determine the strain along the fiber and transverse directions use strain expression (11.11) for $\theta = 60°$. One would have:

$$\begin{Bmatrix} \varepsilon_L \\ \varepsilon_T \\ \frac{\gamma_{LT}}{2} \end{Bmatrix} = \begin{bmatrix} .25 & .75 & .866 \\ .75 & .25 & -.866 \\ -.433 & .433 & -.5 \end{bmatrix} \begin{Bmatrix} -400 \\ 700 \\ -225 \end{Bmatrix} = \begin{Bmatrix} 230 \\ 70 \\ 589 \end{Bmatrix} \text{ thus,}$$

$$\varepsilon_L = 230\mu\varepsilon, \quad \varepsilon_T = 70\mu\varepsilon \text{ and } \gamma_{LT} = 1178\mu\varepsilon$$

Using expression (11.4), one can determine the stresses at the longitudinal and transverse directions. Before that the Q values have to be determined as follows:

$$\text{Note: } v_{LT}E_T = v_{TL}E_L, \text{ then } \frac{v_{LT}E_T}{E_L} = v_{TL} = .02$$

$$Q11 = \frac{E_L}{1 - v_{LT}v_{TL}} = \frac{20E6}{1 - 0.3(0.02)} = 20.12E6$$

$$Q22 = \frac{E_T}{1 - v_{LT}v_{TL}} = \frac{1.3E6}{1 - 0.3(0.02)} = 1.31E6$$

$$Q12 = \frac{v_{LT}E_T}{1 - v_{LT}v_{TL}} = \frac{0.3(1.3E6)}{1 - 0.3(0.02)} = 0.392E6$$

$$Q66 = G_{LT} = 1.03E6$$

Thus, one would have

$$\begin{Bmatrix} \sigma_L \\ \sigma_T \\ \tau_{LT} \end{Bmatrix} = \begin{bmatrix} 20.12E6 & 0.392E6 & 0 \\ 0.392E6 & 1.31E6 & 0 \\ 0 & 0 & 1.03E6 \end{bmatrix} \begin{Bmatrix} 230 \times 10^{-6} \\ 70 \times 10^{-6} \\ 1178 \times 10^{-6} \end{Bmatrix} = \begin{Bmatrix} 4655 \\ 181 \\ 1213 \end{Bmatrix} \text{ thus,}$$

$$\sigma_L = 4655 \text{ psi}, \quad \sigma_T = 181 \text{ psi and } \tau_{LT} = 1213 \text{ psi}$$

Using expression (11.9), one can determine the normal and shear stresses in the x and y direction, as follows:

$$\begin{Bmatrix} \sigma_x \\ \sigma_y \\ \tau_{xy} \end{Bmatrix} = \begin{bmatrix} 0.25 & 0.75 & -0.866 \\ 0.75 & 0.25 & 0.866 \\ 0.433 & -0.433 & -0.5 \end{bmatrix} \begin{Bmatrix} 4655 \\ 181 \\ 1213 \end{Bmatrix} = \begin{Bmatrix} 249 \\ 4586 \\ 1331 \end{Bmatrix}$$

$$\sigma_x = 249 \text{ psi}, \quad \sigma_y = 4586 \text{ psi and } \tau_{xy} = 1331 \text{ psi}$$

11.4 ANALYSIS OF LAMINATE

Assume a single layer of material undergoing bending, as shown in Figure 11.3. Let u, v and w represent the displacements in x, y and z directions, respectively.

After bending, the displacement of the layer is a combination of the mid-plane displacement and the angular displacement, denoted as follows:

$$u = u_0 - z\frac{\partial w_0}{\partial x}$$
$$v = v_0 - z\frac{\partial w_0}{\partial y}$$

(11.21)

where u_0, v_0 and w_0 are the mid-plane displacements.

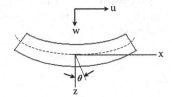

FIGURE 11.3   Single layer bending of material.

The strains are defined as

$$\varepsilon_x = \frac{\partial u}{\partial x} = \frac{\partial u_0}{\partial x} - z\frac{\partial^2 w_0}{\partial x^2} = \varepsilon_x^0 + zk_x$$

(11.22)

$$\varepsilon_x = \frac{\partial v}{\partial y} = \frac{\partial v_0}{\partial y} - z\frac{\partial^2 w_0}{\partial y^2} = \varepsilon_y^0 + zk_y$$

(11.23)

$$\gamma_{xy} = \frac{\partial u}{\partial y} + \frac{\partial v}{\partial x} = \frac{\partial u_0}{\partial y} + \frac{\partial v_0}{\partial x} - 2z\frac{\partial^2 w_0}{\partial x \partial y} = \gamma_{xy}^0 + zk_{xy}$$

(11.24)

where ε_x^0, ε_y^0 and γ_{xy}^0 are the mid-plane strains and k_x, k_y and k_{xy} are the plate curvatures.

These expression can be rewritten in terms of the mid-plane strains and plate curvatures in vector form.

$$\begin{Bmatrix} \varepsilon_x \\ \varepsilon_y \\ \gamma_{xy} \end{Bmatrix} = \begin{Bmatrix} \varepsilon_x^0 \\ \varepsilon_y^0 \\ \gamma_{xy}^0 \end{Bmatrix} + z\begin{Bmatrix} k_x \\ k_y \\ k_{xy} \end{Bmatrix}$$

(11.25)

Substitute the above expression (11.25) into expression (11.13), and the stress-strain relationship for the single layer, k, can be written as

$$\begin{Bmatrix} \sigma_x \\ \sigma_y \\ \tau_{xy} \end{Bmatrix}_k = \begin{bmatrix} \overline{Q11} & \overline{Q12} & \overline{Q16} \\ \overline{Q12} & \overline{Q22} & \overline{Q26} \\ \overline{Q16} & \overline{Q26} & \overline{Q66} \end{bmatrix}_k \begin{Bmatrix} \varepsilon_x \\ \varepsilon_y \\ \gamma_{xy} \end{Bmatrix}$$

(11.26)

or

$$
\left\{ \begin{array}{c} \sigma_x \\ \sigma_y \\ \tau_{xy} \end{array} \right\}_k = \begin{bmatrix} \overline{Q11} & \overline{Q12} & \overline{Q16} \\ \overline{Q12} & \overline{Q22} & \overline{Q26} \\ \overline{Q16} & \overline{Q26} & \overline{Q66} \end{bmatrix}_k \left\{ \begin{array}{c} \varepsilon_x^0 \\ \varepsilon_y^0 \\ \gamma_{xy}^0 \end{array} \right\} + z \begin{bmatrix} \overline{Q11} & \overline{Q12} & \overline{Q16} \\ \overline{Q12} & \overline{Q22} & \overline{Q26} \\ \overline{Q16} & \overline{Q26} & \overline{Q66} \end{bmatrix}_k \left\{ \begin{array}{c} k_x \\ k_y \\ k_{xy} \end{array} \right\}
$$

$$(11.27)$$

Now, consider a laminate with the height of h, made out of n single layers. The resultant forces and the moments acting on the laminate are as follows:

$$
\left\{ \begin{array}{c} N_x \\ N_y \\ N_{xy} \end{array} \right\} = \int_{-h/2}^{h/2} \left\{ \begin{array}{c} \sigma_x \\ \sigma_y \\ \tau_{xy} \end{array} \right\} dz = \sum_{k=1}^{n} \int_{h_{k-1}}^{h_k} \left\{ \begin{array}{c} \sigma_x \\ \sigma_y \\ \tau_{xy} \end{array} \right\} dz
$$

$$(11.28)$$

$$
\left\{ \begin{array}{c} M_x \\ M_y \\ M_{xy} \end{array} \right\} = \int_{-h/2}^{h/2} \left\{ \begin{array}{c} \sigma_x \\ \sigma_y \\ \tau_{xy} \end{array} \right\} z\,dz = \sum_{k=1}^{n} \int_{h_{k-1}}^{h_k} \left\{ \begin{array}{c} \sigma_x \\ \sigma_y \\ \tau_{xy} \end{array} \right\} z\,dz
$$

$$(11.29)$$

where $\{N\}$ is the resultant forces (force per unit length) and $\{M\}$ is the resultant moments (moment per unit length) on the laminate.

Now, substituting expression (11.27) into equations (11.28) and (11.29) one would have

$$
\left\{ \begin{array}{c} N_x \\ N_y \\ N_{xy} \end{array} \right\} = \left[\sum_{k=1}^{n} \begin{bmatrix} \overline{Q11} & \overline{Q12} & \overline{Q16} \\ \overline{Q12} & \overline{Q22} & \overline{Q26} \\ \overline{Q16} & \overline{Q26} & \overline{Q66} \end{bmatrix}_k \int_{h_{k-1}}^{h_k} dz \right] \left\{ \begin{array}{c} \varepsilon_x^0 \\ \varepsilon_y^0 \\ \gamma_{xy}^0 \end{array} \right\}
$$

$$
+ \left[\sum_{k=1}^{n} \begin{bmatrix} \overline{Q11} & \overline{Q12} & \overline{Q16} \\ \overline{Q12} & \overline{Q22} & \overline{Q26} \\ \overline{Q16} & \overline{Q26} & \overline{Q66} \end{bmatrix}_k \int_{h_{k-1}}^{h_k} z\,dz \right] \left\{ \begin{array}{c} k_x \\ k_y \\ k_{xy} \end{array} \right\} \quad (11.30)
$$

$$
\left\{ \begin{array}{c} M_x \\ M_y \\ M_{xy} \end{array} \right\} = \left[\sum_{k=1}^{n} \begin{bmatrix} \overline{Q11} & \overline{Q12} & \overline{Q16} \\ \overline{Q12} & \overline{Q22} & \overline{Q26} \\ \overline{Q16} & \overline{Q26} & \overline{Q66} \end{bmatrix}_k \int_{h_{k-1}}^{h_k} z\,dz \right] \left\{ \begin{array}{c} \varepsilon_x^0 \\ \varepsilon_y^0 \\ \gamma_{xy}^0 \end{array} \right\}
$$

$$
+ \left[\sum_{k=1}^{n} \begin{bmatrix} \overline{Q11} & \overline{Q12} & \overline{Q16} \\ \overline{Q12} & \overline{Q22} & \overline{Q26} \\ \overline{Q16} & \overline{Q26} & \overline{Q66} \end{bmatrix}_k \int_{h_{k-1}}^{h_k} z^2\,dz \right] \left\{ \begin{array}{c} k_x \\ k_y \\ k_{xy} \end{array} \right\} \quad (11.31)
$$

Rewriting expressions (11.30) and (11.31) in another form results

$$\begin{Bmatrix} N_x \\ N_y \\ N_{xy} \end{Bmatrix} = \begin{bmatrix} A11 & A12 & A16 \\ A12 & A22 & A26 \\ A16 & A26 & A66 \end{bmatrix} \begin{Bmatrix} \varepsilon_x^0 \\ \varepsilon_y^0 \\ \gamma_{xy}^0 \end{Bmatrix} + \begin{bmatrix} B11 & B12 & B16 \\ B12 & B22 & B26 \\ B16 & B26 & B66 \end{bmatrix} \begin{Bmatrix} k_x \\ k_y \\ k_{xy} \end{Bmatrix} \quad (11.32)$$

$$\begin{Bmatrix} M_x \\ M_y \\ M_{xy} \end{Bmatrix} = \begin{bmatrix} B11 & B12 & B16 \\ B12 & B22 & B26 \\ B16 & B26 & B66 \end{bmatrix} \begin{Bmatrix} \varepsilon_x^0 \\ \varepsilon_y^0 \\ \gamma_{xy}^0 \end{Bmatrix} + \begin{bmatrix} D11 & D12 & D16 \\ D12 & D22 & D26 \\ D16 & D26 & D66 \end{bmatrix} \begin{Bmatrix} k_x \\ k_y \\ k_{xy} \end{Bmatrix} \quad (11.33)$$

where the $[A]$, $[B]$ and $[D]$ matrices are known as the extensional stiffness matrix, coupling stiffness matrix and bending stiffness matrix, respectively.

The components of these matrices are given by using the laminate lay-up, shown by Figure 11.4:

$$A_{ij} = \sum_{k=1}^{n} (\overline{Q}_{ij})_k (h_k - h_{k-1}) \quad (11.34)$$

$$B_{ij} = \frac{1}{2} \sum_{k=1}^{n} (\overline{Q}_{ij})_k (h_k^2 - h_{k-1}^2) \quad (11.35)$$

$$D_{ij} = \frac{1}{3} \sum_{k=1}^{n} (\overline{Q}_{ij})_k (h_k^3 - h_{k-1}^3) \quad (11.36)$$

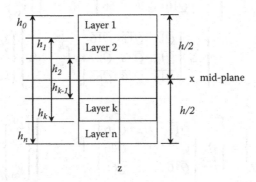

FIGURE 11.4   The laminate lay-up.

Expressions (11.32) and (11.33) can be rewritten to give the strain and curvature results based on the applied resultant loading as follows:

$$
\begin{Bmatrix} \varepsilon_x^0 \\ \varepsilon_y^0 \\ \gamma_{xy}^0 \end{Bmatrix} = [[\alpha] - [\beta][\delta^{-1}][\kappa]] \begin{Bmatrix} N_x \\ N_y \\ N_{xy} \end{Bmatrix} + [\beta][\delta^{-1}] \begin{Bmatrix} M_x \\ M_y \\ M_{xy} \end{Bmatrix} \tag{11.37}
$$

and

$$
\begin{Bmatrix} k_x \\ k_y \\ k_{xy} \end{Bmatrix} = [\delta^{-1}] \begin{Bmatrix} M_x \\ M_y \\ M_{xy} \end{Bmatrix} - [\delta^{-1}][\kappa] \begin{Bmatrix} N_x \\ N_y \\ N_{xy} \end{Bmatrix} \tag{11.38}
$$

where

$$
\begin{aligned}
[\alpha] &= [A^{-1}] \\
[\beta] &= -[A^{-1}][B] \\
[\kappa] &= [B][A^{-1}] \\
[\delta] &= [D] - [B][A^{-1}][B]
\end{aligned} \tag{11.39}
$$

Example 11.2: Consider the laminate lay-up of (0/±45/90) under the loading as shown by the following figure. The lamina properties are $E_L = 20 \times 10^6$ psi, $E_T = 1.3 \times 10^6$ psi, $G_{LT} = 1.03 \times 10^6$ psi, $\upsilon_{LT} = 0.3$ for all layers. Determine the strains in the x and y direction. Also, determine the strains and stresses along the longitudinal and transverse directions for each ply layer. Assume each layer is 0.05 in. thick.

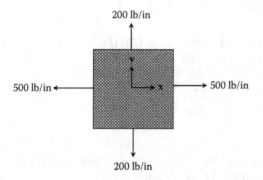

Determine the stiffness matrix [Q] by using expressions (11.5) through (11.8):

$$
Q11 = \frac{E_L}{1 - \upsilon_{LT}\upsilon_{TL}}, \quad Q22 = \frac{E_T}{1 - \upsilon_{LT}\upsilon_{TL}}, \quad Q12 = \frac{\upsilon_{LT}E_T}{1 - \upsilon_{LT}\upsilon_{TL}}
$$

$$
= \frac{\upsilon_{TL}E_L}{1 - \upsilon_{LT}\upsilon_{TL}}, \quad Q66 = G_{LT}
$$

$$[Q] = \begin{bmatrix} 20.12E6 & 0.3923E6 & 0 \\ 0.3923E6 & 1.308E6 & 0 \\ 0 & 0 & 1.03E6 \end{bmatrix}$$

Determine the stiffness matrices for each angled layer (0/45/-45/90) using expressions (11.14) through (11.19):

$$\overline{Q11} = Q11c^4\theta + Q22s^4\theta + 2(Q12 + 2Q66)s^2\theta c^2\theta$$

$$\overline{Q22} = Q22c^4\theta + Q11s^4\theta + 2(Q12 + 2Q66)s^2\theta c^2\theta$$

$$\overline{Q12} = (Q11 + Q22 - 4Q66)s^2\theta c^2\theta + Q12(s^4\theta + c^4\theta)$$

$$\overline{Q66} = (Q11 + Q22 - 2Q12 - 2Q66)s^2\theta c^2\theta + Q66(s^4\theta + c^4\theta)$$

$$\overline{Q16} = (Q11 - Q12 - 2Q66)c^3\theta s\theta - (Q22 - Q12 - 2Q66)c\theta s^3\theta)$$

$$\overline{Q26} = (Q11 - Q12 - 2Q66)c\theta s^3\theta - (Q22 - Q12 - 2Q66)c^3\theta s\theta)$$

$$[\overline{Q}]_0 = \begin{bmatrix} 20.12E6 & 0.3923E6 & 0 \\ 0.3923E6 & 1.308E6 & 0 \\ 0 & 0 & 1.03E6 \end{bmatrix}$$

$$[\overline{Q}]_{45} = \begin{bmatrix} 6.582E6 & 4.522E6 & 4.703E6 \\ 4.522E6 & 6.582E6 & 4.703E6 \\ 4.703E6 & 4.703E6 & 5.160E6 \end{bmatrix}$$

$$[\overline{Q}]_{-45} = \begin{bmatrix} 6.582E6 & 4.522E6 & -4.703E6 \\ 4.522E6 & 6.582E6 & -4.703E6 \\ -4.703E6 & -4.703E6 & 5.160E6 \end{bmatrix}$$

$$[\overline{Q}]_{90} = \begin{bmatrix} 1.308E6 & 0.3923E6 & 0 \\ 0.3923E6 & 20.17E6 & 0 \\ 0 & 0 & 1.03E6 \end{bmatrix}$$

Apply the following lay-up schematic, to calculate the elements of the [A], [B] and [D] matrices using equations (11.34) through (11.36):

$$Aij = \sum_{k=1}^{n} (\overline{Q}ij)_k (h_k - h_{k-1}), \quad Bij = \frac{1}{2} \sum_{k=1}^{n} (\overline{Q}ij)_k (h_k^2 - h_{k-1}^2),$$

$$Dij = \frac{1}{3} \sum_{k=1}^{n} (\overline{Q}ij)_k (h_k^3 - h_{k-1}^3)$$

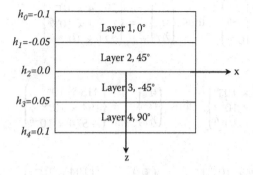

$$[A] = \begin{bmatrix} 1.73E6 & 4.915E5 & 0 \\ 4.915E5 & 1.73E6 & 0 \\ 0 & 0 & 6.19E5 \end{bmatrix}$$

$$[B] = \begin{bmatrix} -7.054E4 & 0 & -1.176E4 \\ 0 & 7.054E4 & -1.176E4 \\ -1.176E4 & -1.176E4 & 0 \end{bmatrix}$$

$$[D] = \begin{bmatrix} 6.798E3 & 6.057E2 & 0 \\ 6.057E2 & 6.798E3 & 0 \\ 0 & 0 & 1.031E3 \end{bmatrix}$$

Using expression (11.37) calculate the laminate strains:

$$\begin{Bmatrix} \varepsilon_x^0 \\ \varepsilon_y^0 \\ \gamma_{xy}^0 \end{Bmatrix} = [[\alpha] - [\beta][\delta^{-1}][\kappa]] \begin{Bmatrix} 500 \\ 200 \\ 0 \end{Bmatrix} + [\beta][\delta^{-1}] \begin{Bmatrix} 0 \\ 0 \\ 0 \end{Bmatrix},$$

$$\text{thus} \begin{Bmatrix} \varepsilon_x^0 \\ \varepsilon_y^0 \\ \gamma_{xy}^0 \end{Bmatrix} = \begin{Bmatrix} 591.4 \times 10^{-6} \\ 17.64 \times 10^{-6} \\ 110.5 \times 10^{-6} \end{Bmatrix}$$

Transform the laminate strains to the strains at each layer (0/45/-45/90) using expression (11.11):

$$\begin{Bmatrix} \varepsilon_L \\ \varepsilon_T \\ \frac{\gamma_{LT}}{2} \end{Bmatrix} = \begin{bmatrix} c^2\theta & s^2\theta & 2s\theta c\theta \\ s^2\theta & c^2\theta & -2s\theta c\theta \\ -s\theta c\theta & s\theta c\theta & c^2\theta - s^2\theta \end{bmatrix} \begin{Bmatrix} \varepsilon_x \\ \varepsilon_y \\ \frac{\gamma_{xy}}{2} \end{Bmatrix}$$

$$\begin{Bmatrix} \varepsilon_L \\ \varepsilon_T \\ \frac{\gamma_{LT}}{2} \end{Bmatrix}_{0°} = \begin{bmatrix} 1 & 0 & 0 \\ 0 & 1 & 0 \\ 0 & 0 & 1 \end{bmatrix} \begin{Bmatrix} 591.4 \times 10^{-6} \\ 17.64 \times 10^{-6} \\ 110.5 \times 10^{-6} \end{Bmatrix}, \quad \text{thus} \quad \begin{Bmatrix} \varepsilon_L \\ \varepsilon_T \\ \gamma_{LT} \end{Bmatrix}_{0°} = \begin{Bmatrix} 591.4 \times 10^{-6} \\ 17.64 \times 10^{-6} \\ 221 \times 10^{-6} \end{Bmatrix}$$

$$\begin{Bmatrix} \varepsilon_L \\ \varepsilon_T \\ \frac{\gamma_{LT}}{2} \end{Bmatrix}_{45°} = \begin{bmatrix} .5 & .5 & 1 \\ .5 & .5 & -1 \\ -.5 & .5 & 0 \end{bmatrix} \begin{Bmatrix} 591.4 \times 10^{-6} \\ 17.64 \times 10^{-6} \\ 110.5 \times 10^{-6} \end{Bmatrix}, \quad \text{thus} \quad \begin{Bmatrix} \varepsilon_L \\ \varepsilon_T \\ \gamma_{LT} \end{Bmatrix}_{45°} = \begin{Bmatrix} 415 \times 10^{-6} \\ 194 \times 10^{-6} \\ -574 \times 10^{-6} \end{Bmatrix}$$

$$\begin{Bmatrix} \varepsilon_L \\ \varepsilon_T \\ \frac{\gamma_{LT}}{2} \end{Bmatrix}_{-45°} = \begin{bmatrix} .5 & .5 & -1 \\ .5 & .5 & 1 \\ .5 & -.5 & 0 \end{bmatrix} \begin{Bmatrix} 591.4 \times 10^{-6} \\ 17.64 \times 10^{-6} \\ 110.5 \times 10^{-6} \end{Bmatrix}, \quad \text{thus} \quad \begin{Bmatrix} \varepsilon_L \\ \varepsilon_T \\ \gamma_{LT} \end{Bmatrix}_{-45°} = \begin{Bmatrix} 194 \times 10^{-6} \\ 415 \times 10^{-6} \\ 574 \times 10^{-6} \end{Bmatrix}$$

$$\begin{Bmatrix} \varepsilon_L \\ \varepsilon_T \\ \frac{\gamma_{LT}}{2} \end{Bmatrix}_{90°} = \begin{bmatrix} 0 & 1 & 0 \\ 1 & 0 & 0 \\ 0 & 0 & -1 \end{bmatrix} \begin{Bmatrix} 591.4 \times 10^{-6} \\ 17.64 \times 10^{-6} \\ 110.5 \times 10^{-6} \end{Bmatrix}, \quad \text{thus} \quad \begin{Bmatrix} \varepsilon_L \\ \varepsilon_T \\ \gamma_{LT} \end{Bmatrix}_{90°} = \begin{Bmatrix} 17.64 \times 10^{-6} \\ 591.4 \times 10^{-6} \\ -221 \times 10^{-6} \end{Bmatrix}$$

Using expression (11.4) determine the stresses in each ply layer.

$$\begin{Bmatrix} \sigma_L \\ \sigma_T \\ \tau_{LT} \end{Bmatrix} = \begin{bmatrix} Q11 & Q12 & 0 \\ Q12 & Q22 & 0 \\ 0 & 0 & Q66 \end{bmatrix} \begin{Bmatrix} \varepsilon_L \\ \varepsilon_T \\ \gamma_{LT} \end{Bmatrix}$$

$$\begin{Bmatrix} \sigma_L \\ \sigma_T \\ \tau_{LT} \end{Bmatrix}_{0°} = \begin{bmatrix} 20.12E6 & 0.392E6 & 0 \\ 0.392E6 & 1.31E6 & 0 \\ 0 & 0 & 1.03E6 \end{bmatrix} \begin{Bmatrix} 591.4 \times 10^{-6} \\ 17.64 \times 10^{-6} \\ 221 \times 10^{-6} \end{Bmatrix}_{0°} = \begin{Bmatrix} 11910 \\ 255 \\ 228 \end{Bmatrix} \text{ psi}$$

$$\begin{Bmatrix} \sigma_L \\ \sigma_T \\ \tau_{LT} \end{Bmatrix}_{45°} = \begin{bmatrix} 20.12E6 & 0.392E6 & 0 \\ 0.392E6 & 1.31E6 & 0 \\ 0 & 0 & 1.03E6 \end{bmatrix} \begin{Bmatrix} 415 \times 10^{-6} \\ 194 \times 10^{-6} \\ -574 \times 10^{-6} \end{Bmatrix}_{45°} = \begin{Bmatrix} 8426 \\ 417 \\ -591 \end{Bmatrix} \text{ psi}$$

$$\begin{Bmatrix} \sigma_L \\ \sigma_T \\ \tau_{LT} \end{Bmatrix}_{-45°} = \begin{bmatrix} 20.12E6 & 0.392E6 & 0 \\ 0.392E6 & 1.31E6 & 0 \\ 0 & 0 & 1.03E6 \end{bmatrix} \begin{Bmatrix} 194 \times 10^{-6} \\ 415 \times 10^{-6} \\ 574 \times 10^{-6} \end{Bmatrix}_{-45°} = \begin{Bmatrix} 4066 \\ 620 \\ 591 \end{Bmatrix} \text{ psi}$$

$$\begin{Bmatrix} \sigma_L \\ \sigma_T \\ \tau_{LT} \end{Bmatrix}_{90°} = \begin{bmatrix} 20.12E6 & 0.392E6 & 0 \\ 0.392E6 & 1.31E6 & 0 \\ 0 & 0 & 1.03E6 \end{bmatrix} \begin{Bmatrix} 17.64 \times 10^{-6} \\ 591.4 \times 10^{-6} \\ -221 \times 10^{-6} \end{Bmatrix}_{90°} = \begin{Bmatrix} 587 \\ 782 \\ -228 \end{Bmatrix} \text{ psi}$$

11.5 EFFECTIVE MODULUS OF THE LAMINATE

Once a laminate is designed and built up then the effective properties of the laminate can be used to analyze the laminate state of stress and strains as well. The effective properties of a laminate can be evaluated by use of the $[A]$ matrix as follows:

$$E_{x(eff)} = \frac{1}{h}\left(\frac{A11A22 - A12^2}{A22}\right) \tag{11.40}$$

$$E_{y(eff)} = \frac{1}{h}\left(\frac{A11A22 - A12^2}{A11}\right) \tag{11.41}$$

$$G_{xy(eff)} = \frac{1}{h}A66 \tag{11.42}$$

$$v_{xy(eff)} = \frac{A12}{A22} \tag{11.43}$$

There are several laminate lay-ups that produce effective modulus properties that are similar in the in-plane directions and mimic an isotropic material. These laminates are referred to as the quasiisotropic laminates.

The laminate shown in example 11.2 (0/45/-45/90) is a quasiisotropic laminate with effective properties given as

$$E_{x(eff)} = \frac{1}{0.2}\left(\frac{1.73E6(1.73E6) - (4.915E5)^2}{1.73E6}\right) = 7.951 \times 10^6 \, psi$$

$$E_{y(eff)} = \frac{1}{0.2}\left(\frac{1.73E6(1.73E6) - (4.915E5)^2}{1.73E6}\right) = 7.951 \times 10^6 \, psi$$

$$G_{xy(eff)} = \frac{1}{0.2}(6.19E5) = 3.095 \times 10^6 \, psi$$

$$v_{xy(eff)} = \frac{4.915E5}{1.73E6} = 0.28$$

One can observe that the E_x and E_y are equal. Another quasiisotropic laminate lay-up is the (0/60/-60).

Problems

1. For an IM6/Epoxy lamina oriented at 45° from the x-axis, the strains levels are, $\varepsilon_x = 800 \, \mu\varepsilon$, $\varepsilon_y = 950 \, \mu\varepsilon$ and $\gamma_{xy} = -500 \, \mu\varepsilon$. Determine the

normal stress components in the x and y direction and the shear stress. Also determine the stresses and strains along the fiber and transverse matrix directions.

2. For an E-glass/epoxy lamina oriented at 30° from the x-axis, the stress levels are $\sigma_x = 1500\ psi$, $\sigma_y = 200\ psi$. Determine the in-plane strains resulting from this loading.

3. Consider a laminate lay-up of (0/60/-60) with each layer being 0.03 in. thick. This laminate is made of AS/3501 plies. Determine the [A], [B] and [D] matrices for this build up.

4. Consider a laminate lay-up of (45/-45/90/90/-45/45) with each layer being 0.05 in. thick. This laminate is made of IM6/Epoxy plies. Determine the [A], [B] and [D] matrices for this build up.

5. Consider a laminate lay-up of (0/90/90/0) with each layer being 0.08 in. thick. This laminate is made of S-glass/Epoxy plies. Determine the [A], [B] and [D] matrices for this build up.

6. For the problem 4 above under x-direction loading of 800 lb/in and y-direction loading of 350 lb/in., determine the strains in the x and y direction.

7. Consider a laminate lay-up of (0/45/90/90/45/0) with each layer being 0.05 in. thick. This laminate is made of IM6/Epoxy plies. The applied loading in the x-direction is 1000 lb/in., and the loading in the y-direction is 1300 lb/in. Determine the strains in the x and y direction. Also, determine the strains and stresses along the longitudinal and transverse directions for each ply layer.

8. Repeat problem 7 above with laminate lay-up of (0/45/30/30/45/0).

REFERENCES

Agarwal, B. D., Broutman, L.J., *Analysis and Performance of Fiber Composites*, 1990. New York: John Wiley & Sons.

Gurdal, Z., Haftka, R.T., Hajela, P., *Design and Optimization of Laminated Composite Materials*, 1999. New York: John Wiley & Sons.

Ugural, A.C., Fenster, S.K., *Advanced Strength and Applied Elasticity*, 1995. New Jersey: Prentice Hall.

12 Fasteners and Joint Connections

12.1 INTRODUCTION

Fasteners and joints are integral parts of the design of structures. Thus, stress analysts must learn about stress analysis of fasteners and joints. There are two major types of connections: 1) bolted or riveted connections and 2) welded connections. This chapter illustrates fastener-joint failure modes of the first type of connection and failure modes of the second type. Failure stress calculations are shown for each type of failure. Furthermore, fastener analysis under eccentric loading of the joints is shown. For this analysis, basic knowledge of vector mechanics is necessary.

12.2 FASTENER-CONNECTION FAILURE

Bolted and riveted connection failure is categorized into two sections: 1) fastener failure and 2) joint failure. The fastener failure modes are normally due to shear and bending of the fasteners (bolts and rivets). Figure 12.1 illustrates a fastener connection under shear loading.

FIGURE 12.1 Fastener under shear loading.

The shear stress on the fastener is calculated as

$$\tau = \frac{F}{A_s} \tag{12.1}$$

where A_s is the cross-sectional area of the fastener with diameter D, ($A_s = \pi \frac{D^2}{4}$). Figure 12.2 illustrates a fastener under combined bending and shear loading.

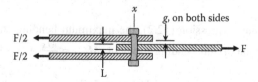

FIGURE 12.2 Fastener under bending and shear loading.

DOI: 10.1201/9781003311218-12

The bending stress on the fastener is calculated as

$$\sigma = \frac{MD}{2I} \tag{12.2}$$

where I is the moment of inertia of the fastener with diameter D, ($I = \pi \frac{D^4}{64}$).

The maximum bending moment (M) and shear (V) are calculated assuming the following load distribution on the fastener shown in Figure 12.3,

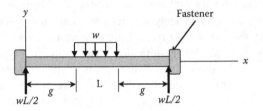

FIGURE 12.3   The fastener load distribution.

whereas the distributed load exerted by the middle lap is $w = \frac{F}{L}$.

Making use of the shear and moment diagrams for this fastener, one would have (Figure 12.4).

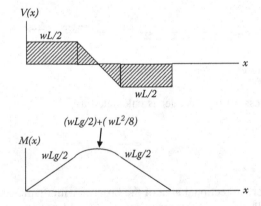

FIGURE 12.4   The shear-moment diagram for the fastener.

Hence, by using expression (12.2), the maximum bending stress on the fastener due to the maximum bending moment, $M_{\max} = \frac{wLg}{2} + \frac{wL^2}{8}$ is

$$\sigma_{\max} = \frac{16wLg + 4wL^2}{\pi D^3} \tag{12.3}$$

or

$$\sigma_{max} = \frac{16Fg + 4FL}{\pi D^3} \qquad (12.4)$$

The maximum shear stress on the fastener due to the maximum shear load, $V_{max} = \frac{wL}{2}$, is

$$\tau = \frac{V_{max} Q}{Ib} \qquad (12.5)$$

where the first moment of area, Q, is: $Q = \frac{2r^3}{3}$ and $I = \frac{\pi r^4}{4}$ and $b = D = 2r$.
Thus, maximum shear stress is

$$\tau = \frac{2wL}{3\pi r^2} \quad \tau = \frac{\frac{wL}{2}}{\frac{\pi r^4}{4}} \frac{2r^3}{3D} = 8wl/3\pi r^2 \quad \tau = \frac{\frac{wL}{2}}{\frac{\pi r^4}{4}b} \frac{2r^3}{3} \qquad (12.6)$$

This equation can be rewritten in terms of load, F, as

$$\tau = \frac{8F}{3\pi D^2} \qquad (12.7)$$

The joint failure modes are due to shear tear-out, tension tear-out and bearing. Figure 12.5 illustrates shear tear-out of the joint under axial loading.

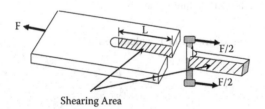

Shearing Area

FIGURE 12.5 The shear tear-out of the joint.

The shear tear-out stress of the joint is calculated as

$$\tau = \frac{F}{2A_{sh}} \qquad (12.8)$$

where A_{sh} is the area of the shear surfaces, $A_{sh} = Lt$.

Figure 12.6 below illustrates the tension tear-out of the joint.

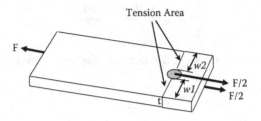

FIGURE 12.6   The tension tear-out of the joint.

The tension tear-out stress of the joint is calculated as

$$\sigma = \frac{F}{A_{tn}} \tag{12.9}$$

where A_{tn} is the tension surface area, $A_{tn} = w1(t) + w2(t)$.

Figure 12.7 illustrates the bearing failure mode of the joint.

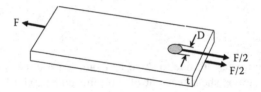

FIGURE 12.7   The bearing failure mode.

The bearing stress of the joint is calculated as

$$\sigma_{br} = \frac{F}{A_{br}} \tag{12.10}$$

where A_{br} is the bearing contact surface of the joint, $A_{br} = Dt$.

Normally a detail analysis carries all of the above failure modes, both for the fasteners and the connecting joints.

12.3   WELDED-CONNECTION FAILURE

The most common type welded connections is the fillet weld connection. Figure 12.8 illustrates the connection and its geometry. The fillet welds transfer the load between the two plates and carry a shear stress in both longitudinal and transverse joint directions.

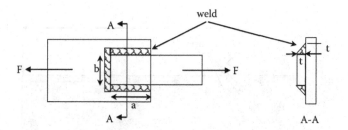

FIGURE 12.8 Typical welded joint connection.

Assume a tensile load of F acting on the joint and the welds having a shearing stress capability of τ_{long} and τ_{trans} in the longitudinal and transverse joint directions, respectively. The following force balance is applicable to the weld connection, which can be used to determine the weld size or the joint force capability depending on what is known to the stress analyst. Note that in this joint there are three sides welded. Depending on the number of sides the joint is welded, the corresponding weld area in the equation below should be adjusted to account for the additional weld areas for this equality to be valid. The areas in this equation are represented by the $(a)(t)$ and $(b)(t)$ terms.

$$F = 2\tau_{long}(a)(t) + \tau_{trans}(b)(t) \qquad (12.11)$$

This is valid for axial tensile load on the joint. If this joint is loaded by a bending moment, the moment should be reduced to a force-couple, and the same force balance should be applied to the joint for weld sizing.

12.4 ECCENTRIC LOADING OF FASTENERS

For a fastener connection, where the loading is applied at a distance, Le from the center of the fastener connection, eccentric loading is developed. Under this condition, there would be a moment developed on the fastener pattern that also introduces a shear loading on each fastener in addition to the simple shear due to the eccentric load. Figures 12.9 and 12.10 illustrate this concept.

The forces F_1, F_2, F_3 and F_4 due to the eccentric moment $M = P(Le)$ are calculated as

$$F_i = \frac{P(Le)}{\sum_{i=1}^{n} r_i^2} r_i \qquad (12.12)$$

where radius r_i is defined as $r_i = \sqrt{x_i^2 + y_i^2}$ and n is the number of fasteners $(n = 4)$.

Now, in order to find the total shear load on each fastener, the resultant load has to be calculated from tload components F_i and (P/n) by utilizing the vector

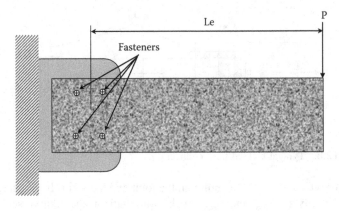

FIGURE 12.9 Eccentric loading of a fastener pattern.

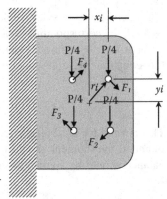

FIGURE 12.10 The load components on the fasteners.

mechanics concepts. The free-body-diagram below in Figure 12.11, illustrates the loads on a typical single fastener. Thus, total shear load on any fastener can easily be determined by the principals of static.

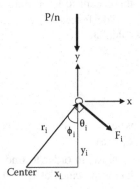

FIGURE 12.11 The free-body-diagram of any fastener.

$$F_{ix} = Cos(\phi_i)F_i \tag{12.13}$$

$$F_{iy} = Sin(\phi_i)F_i \tag{12.14}$$

$$F_{Shear} = \sqrt{\left(\frac{P}{n} + Sin(\phi_i)F_i\right)^2 + \left(Cos(\phi_i)F_i\right)^2} \tag{12.15}$$

Example 12.1: For the bolt pattern shown below, determine the resultant load on each bolt.

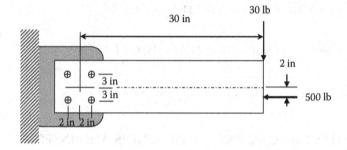

Forces F_1, F_2, F_3 and F_4 due to total eccentric moments, $M = (30)(30) + (500)(2) = 1900$ in-lbf are calculated using equation (12.12):

$$F_i = \frac{M}{\sum_{i=1}^{n} r_i^2} r_i$$

where $r_i = 3.61$ in (a = 2 in. and b = 3 in.)
$F_1 = \frac{1900}{4 \times (3.61^2)} 3.61 = 131.6$ lbf, since the bolt pattern c.g. is right at the center (all r's are equal), then $F_4 = F_3 = F_2 = F_1$
    Free-Body Diagram is:

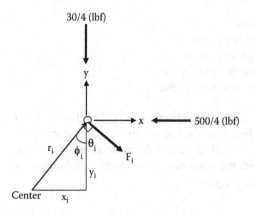

From the free-body-diagram of a fastener shown above for a single general bolt, one has (for maximum possible loading condition consider the lower right hand side bolt in the bolt group, which the load components produced would be in the same direction as the 1000 lbf down and 500 lbf left loads):

$$(\phi_i) = inv\ sin\left(\frac{x_i}{r_i}\right) = inv\ sin\left(\frac{2}{3.61}\right) = 33.64°$$

$$(\theta_i) = 90 - 33.64 = 56.36°$$

$$F_{iy} = Cos(\theta_i)F_i = Cos(56.36)(131.6) = 72.9\ \text{lbf}$$

$$F_{ix} = Sin(\theta_i)F_i = Sin(56.36)(131.6) = 109.6\ \text{lbf}$$

$$F_{Shear} = \sqrt{\left(\frac{30}{4} + 72.9\right)^2 + \left(\frac{500}{4} + 109.6\right)^2} = 248.0\ \text{lbf for all fasteners}$$

12.5   INTER-COUPLING LOADING OF FASTENERS AND INSERTS

In a real-life design consideration, the loads acting on the fasteners and/or the connecting inserts are a combination of tensile (or compressive) and shear loads. For every fastener and insert connection, usually the loads can be reduced to a single combined tensile and shear loading applicable per fastener or insert. For a fastener connection or insert where combined tensile and shear loading is acting as a coupling, the inter-coupling fastener or insert analysis can be done as following to calculate the margins of safety (MS):

$$MS = \frac{1}{\sqrt{\left(\frac{FS(Pt)}{Ft}\right)^2 + \left(\frac{FS(Ps)}{Fs}\right)^2}} - 1 \qquad (12.16)$$

where FS is the factor of safety used:
   Pt is the tensile load on the fastener or insert;
   Ps is the shear load on the fastener or insert;
   Ft is the tensile allowable for the fastener or inserts;
   Fs is the shear allowable for the fastener or inserts.

Note for a bolt or fastener group, the total load would be reduced to load per fastener in the group according to the c.g. of the bolt group and the application of the loads and moments. After this load reduction and determination of the single tensile and shear loading per fastener the same margin of safety calculations can be used per each bolt or fastener in the group.

12.6   FASTENER PROPERTIES

The fastener strength values are shown for common fasteners used in design both in SI and English units. Tables 12.1 and 12.2 show the English and Si versions of the fasteners, respectively.

TABLE 12.1
The Fastener Properties in English Units

SAE Grade	Diameter d (in.)	Proof Strength (psi)	Yield Strength (psi)	Tensile Strength (psi)
1	¼ thru 1 ½	33	36	60
2	¼ thru ¾	55	57	74
2	Over ¾ to 1 ½	33	36	60
5	¼ thru 1	85	92	120
5	Over 1 to 1 ½	74	81	105
5.2	¼ thru 1	85	92	120
7	¼ thru 1 ½	105	115	133
8	¼ thru 1 ½	120	130	150

Source: Society of Automotive Engineering standard J429K (1979).

TABLE 12.2
The Fastener Properties in SI Units

SAE Grade	Diameter d (mm)	Proof Strength (MPa)	Yield Strength (MPa)	Tensile Strength (MPa)
4.6	5 thru 36	225	240	400
4.8	1.6 thru 16	310	340	420
5.8	5 thru 24	380	420	520
8.8	17 thru 36	600	660	830
9.8	1.6 thru 16	650	720	900
10.9	6 thru 36	830	940	1040
12.9	1.6 thru 36	970	1100	1220

Source: Society of Automotive Engineering standard J1199 (1979).

Problems

1. For the single side fastener joint shown, under a fastener loading of 3600 lb, determine the failure modes of the joint and evaluate the corresponding stresses. If an aluminum 7075 material is used for the joint, would the joint be strong enough?

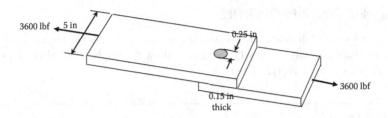

2. For the joint connection shown, determine the maximum stresses on the fastener. Assume two rows of fasteners. The lap thickness, L, is 0.25 in., the gap, g, is 0.1 in. The applied loading, F, is 1200 lbf. The fasteners used are ¼ inch diameter SAE grade 2. Would the fasteners be adequate to hold the joint together?

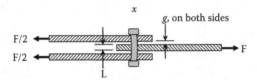

3. For the bolt pattern shown below, determine the resultant load on each bolt. What size bolt do you recommend for this bolt pattern?

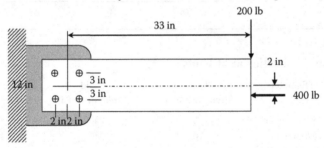

REFERENCES

Beer, F.P., Johnston, E.R., DeWolf, J.T., *Mechanics of Material*, 2002. New York: McGraw Hill.

Ugural, A.C., Fenster, S.K., *Advanced Strength and Applied Elasticity*, 1995. New Jersey: Prentice Hall.

13 MathCAD® Stress Analysis Simulations

13.1 INTRODUCTION

In this chapter, sample MathCAD simulation codes are written to perform various stress analysis calculations and procedures. MathCAD is a commercial symbolic math package tool that can be used to perform various calculations via computer. Basically, the computations are carried out within the package as a series of equations that are typed in as-is. This package contains operational math solvers and tools such as matrix manipulators that are useful for structural analysis. The codes shown in this chapter mimic the calculation analyses that were presented in chapters 1–12 of this book. MathCAD simulation worksheets shown here start with simple section-property calculations, and as the chapters progress, more complex analysis worksheets are presented. Even though each worksheet is designed for a specific analysis procedure, based on the analysis requirements they can be compiled together in any combination for a more comprehensive analysis. It should be noted that in this chapter it is assumed that the reader is familiar with the basic MathCAD operations and formats. Thus, most of the codes are presented without explanation, assuming the reader can input them properly.

Note the square blue boxes appearing within the worksheets are the explanation notes added to ease off the coding within the MathCAD. The 2-D plots are within the capabilities of the MathCAD software and are readily available within the software as tools. Pay careful attention for entering the actual correct units for the starting components you are using since the final derived values would automatically have the correct units if the derivations are correct. This would serve as a good engineering check for your calculations (i.e., F=m a, [N]= [kg] [m/sec^2]).

13.2 SECTION PROPERTY CALCULATIONS

The section properties such as area, moment of inertia and first moment of area can easily be calculated using MathCAD. The following codes are written in MathCAD as sample section property calculations for some of the common structural cross-sections.

® Mathsoft Engineering & Education, Inc.

DOI: 10.1201/9781003311218-13

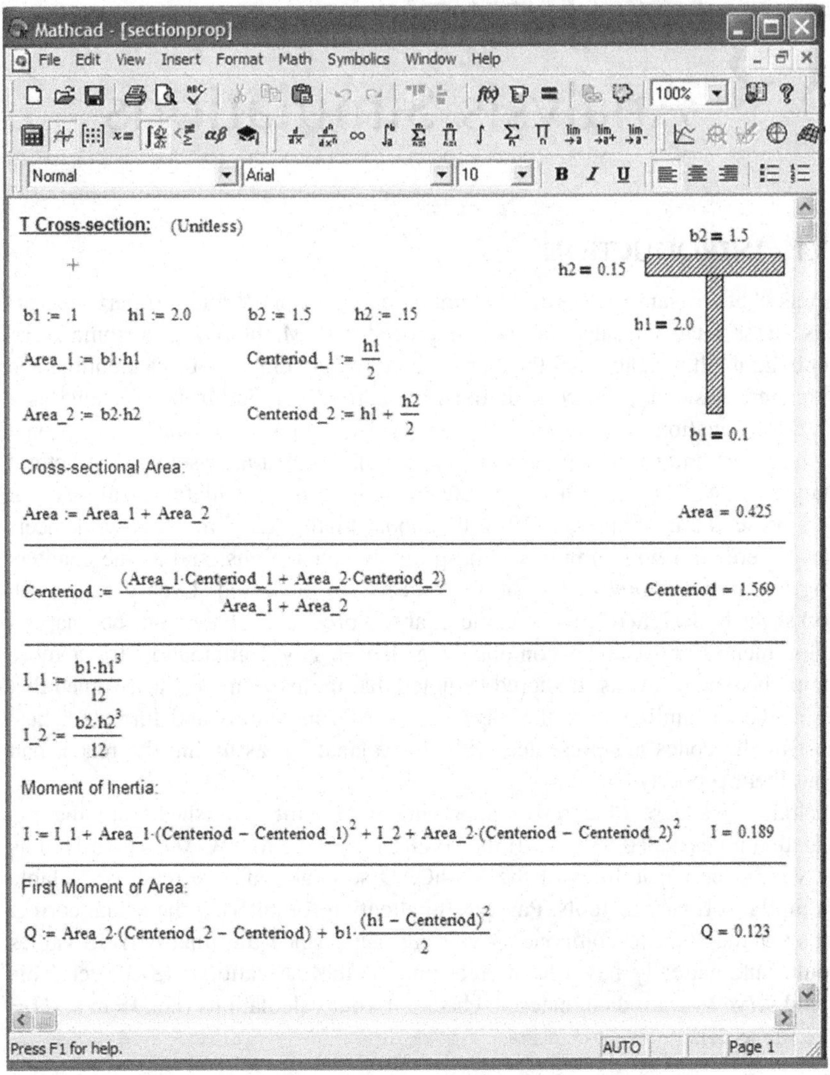
Mathcad - [sectionprop]
File Edit View Insert Format Math Symbolics Window Help
Normal
Arial
10
T Cross-section: (Unitless)
b2 = 1.5
h2 = 0.15
h1 = 2.0
b1 = 0.1
b1 := .1 h1 := 2.0 b2 := 1.5 h2 := .15
Area_1 := b1·h1 Centeriod_1 := h1/2
Area_2 := b2·h2 Centeriod_2 := h1 + h2/2
Cross-sectional Area:
Area := Area_1 + Area_2
Area = 0.425
Centeriod := (Area_1·Centeriod_1 + Area_2·Centeriod_2)/(Area_1 + Area_2)
Centeriod = 1.569
I_1 := b1·h1^3/12
I_2 := b2·h2^3/12
Moment of Inertia:
I := I_1 + Area_1·(Centeriod − Centeriod_1)^2 + I_2 + Area_2·(Centeriod − Centeriod_2)^2
I = 0.189
First Moment of Area:
Q := Area_2·(Centeriod_2 − Centeriod) + b1·(h1 − Centeriod)^2/2
Q = 0.123
Press F1 for help.
AUTO
Page 1

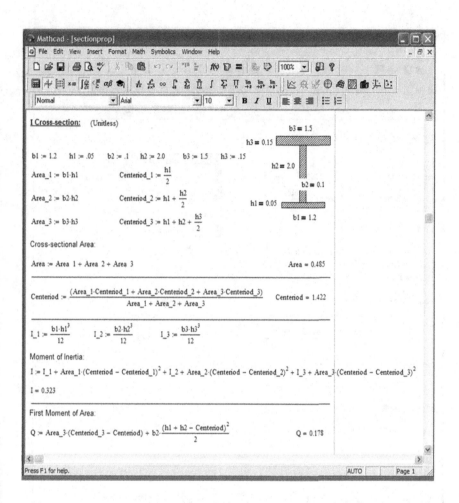

I Cross-section: (Unitless)

$b3 = 1.5$

$h3 = 0.15$

$b1 := 1.2$ $h1 := .05$ $b2 := .1$ $h2 := 2.0$ $b3 := 1.5$ $h3 := .15$

$h2 = 2.0$

$Area\_1 := b1 \cdot h1$ $Centeriod\_1 := \dfrac{h1}{2}$

$b2 = 0.1$

$Area\_2 := b2 \cdot h2$ $Centeriod\_2 := h1 + \dfrac{h2}{2}$

$h1 = 0.05$

$Area\_3 := b3 \cdot h3$ $Centeriod\_3 := h1 + h2 + \dfrac{h3}{2}$

$b1 = 1.2$

Cross-sectional Area:

$Area := Area\_1 + Area\_2 + Area\_3$ $\qquad Area = 0.485$

$Centeriod := \dfrac{(Area\_1 \cdot Centeriod\_1 + Area\_2 \cdot Centeriod\_2 + Area\_3 \cdot Centeriod\_3)}{Area\_1 + Area\_2 + Area\_3}$ $\quad Centeriod = 1.422$

$I\_1 := \dfrac{b1 \cdot h1^3}{12}$ $I\_2 := \dfrac{b2 \cdot h2^3}{12}$ $I\_3 := \dfrac{b3 \cdot h3^3}{12}$

Moment of Inertia:

$I := I\_1 + Area\_1 \cdot (Centeriod - Centeriod\_1)^2 + I\_2 + Area\_2 \cdot (Centeriod - Centeriod\_2)^2 + I\_3 + Area\_3 \cdot (Centeriod - Centeriod\_3)^2$

$I = 0.323$

First Moment of Area:

$Q := Area\_3 \cdot (Centeriod\_3 - Centeriod) + b2 \cdot \dfrac{(h1 + h2 - Centeriod)^2}{2}$ $\qquad Q = 0.178$

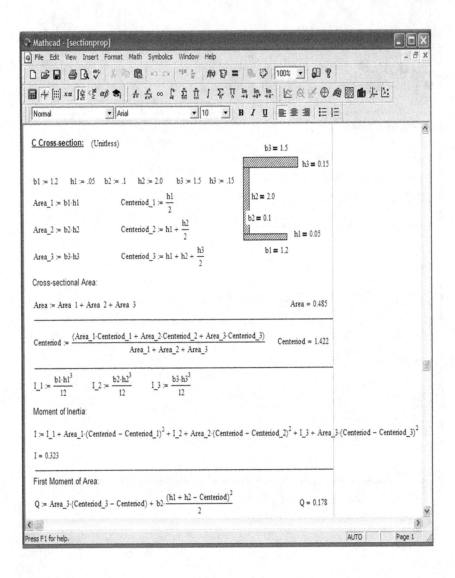

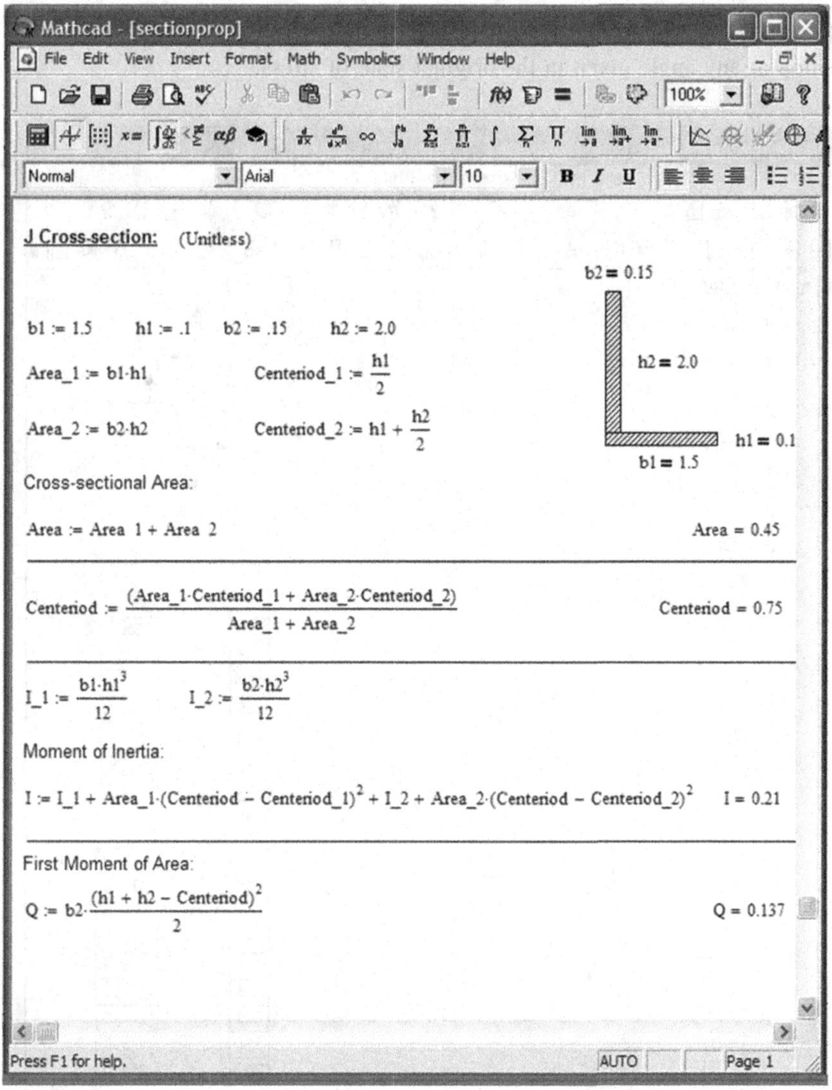
Mathcad - [sectionprop]
File Edit View Insert Format Math Symbolics Window Help
Normal
Arial
10
J Cross-section: (Unitless)
b2 = 0.15
h2 = 2.0
h1 = 0.1
b1 = 1.5
b1 := 1.5 h1 := .1 b2 := .15 h2 := 2.0
Area_1 := b1·h1
Centeriod_1 := h1/2
Area_2 := b2·h2
Centeriod_2 := h1 + h2/2
Cross-sectional Area:
Area := Area_1 + Area_2
Area = 0.45
Centeriod := (Area_1·Centeriod_1 + Area_2·Centeriod_2)/(Area_1 + Area_2)
Centeriod = 0.75
I_1 := b1·h1^3/12 I_2 := b2·h2^3/12
Moment of Inertia:
I := I_1 + Area_1·(Centeriod − Centeriod_1)^2 + I_2 + Area_2·(Centeriod − Centeriod_2)^2
I = 0.21
First Moment of Area:
Q := b2·(h1 + h2 − Centeriod)^2/2
Q = 0.137
Press F1 for help.
AUTO
Page 1

13.3   STRESS TRANSFORMATION AND PRINCIPAL STRESSES

The following MathCAD worksheet was designed to perform stress transformation at any angle given in the original state of stress.

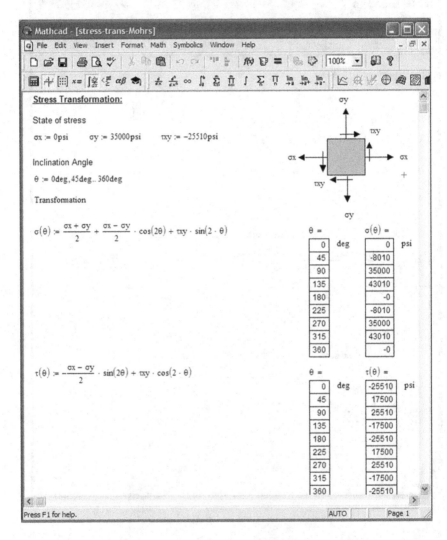

The following MathCAD worksheet was designed to calculate the principal stresses and construct the Mohr's circle. (It should be used concurrently with the previous worksheet.)

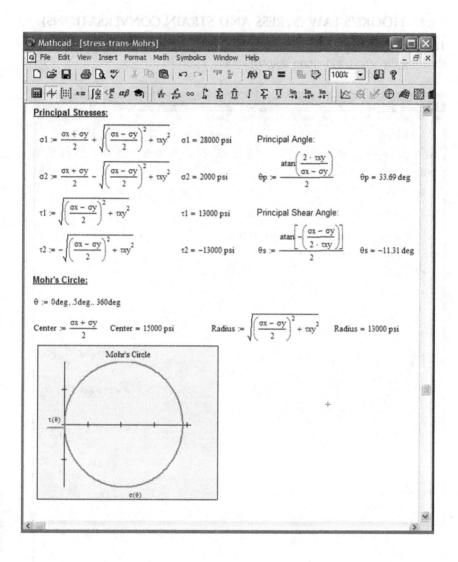

Principal Stresses:

$$\sigma1 := \frac{\sigma x + \sigma y}{2} + \sqrt{\left(\frac{\sigma x - \sigma y}{2}\right)^2 + \tau xy^2} \qquad \sigma1 = 28000\,psi$$

Principal Angle:

$$\sigma2 := \frac{\sigma x + \sigma y}{2} - \sqrt{\left(\frac{\sigma x - \sigma y}{2}\right)^2 + \tau xy^2} \qquad \sigma2 = 2000\,psi \qquad \theta p := \frac{atan\left(\frac{2 \cdot \tau xy}{\sigma x - \sigma y}\right)}{2} \qquad \theta p = 33.69\,deg$$

$$\tau1 := \sqrt{\left(\frac{\sigma x - \sigma y}{2}\right)^2 + \tau xy^2} \qquad \tau1 = 13000\,psi$$

Principal Shear Angle:

$$\tau2 := -\sqrt{\left(\frac{\sigma x - \sigma y}{2}\right)^2 + \tau xy^2} \qquad \tau2 = -13000\,psi \qquad \theta s := \frac{atan\left[-\left(\frac{\sigma x - \sigma y}{2 \cdot \tau xy}\right)\right]}{2} \qquad \theta s = -11.31\,deg$$

Mohr's Circle:

$$\theta := 0deg, .5deg .. 360deg$$

$$Center := \frac{\sigma x + \sigma y}{2} \qquad Center = 15000\,psi \qquad Radius := \sqrt{\left(\frac{\sigma x - \sigma y}{2}\right)^2 + \tau xy^2} \qquad Radius = 13000\,psi$$

Mohr's Circle

$\tau(\theta)$

$\sigma(\theta)$

13.4   HOOKE'S LAW (STRESS AND STRAIN CONVERSATIONS)

The following MathCAD worksheet was designed to illustrate Hooke's law, showing the stress and strain conversations.

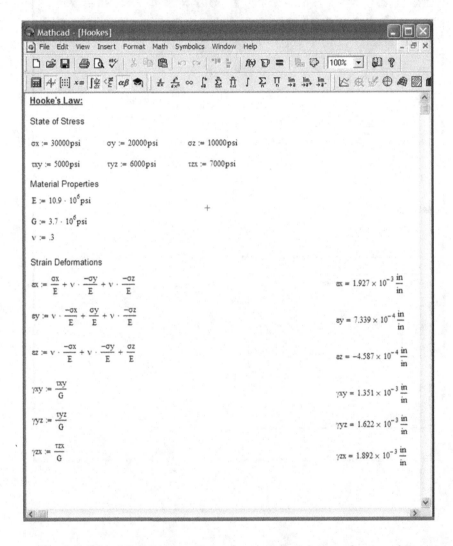

The following MathCAD worksheet was designed to perform strain transformations from the measured strain inputs to the strains in the x, y and shear strain directions.

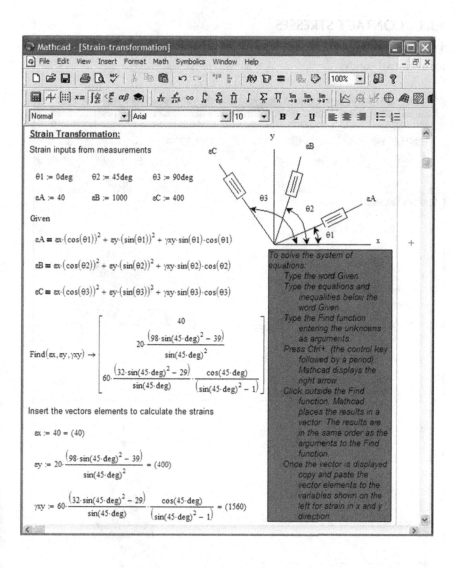

File   Edit   View   Insert   Format   Math   Symbolics   Window   Help

Normal   Arial   10   **B** *I* U

Strain Transformation:

Strain inputs from measurements

$\theta 1 := 0\text{deg}$ $\theta 2 := 45\text{deg}$ $\theta 3 := 90\text{deg}$

$\varepsilon A := 40$ $\varepsilon B := 1000$ $\varepsilon C := 400$

Given

$$\varepsilon A = \varepsilon x \cdot (\cos(\theta 1))^2 + \varepsilon y \cdot (\sin(\theta 1))^2 + \gamma xy \cdot \sin(\theta 1) \cdot \cos(\theta 1)$$

$$\varepsilon B = \varepsilon x \cdot (\cos(\theta 2))^2 + \varepsilon y \cdot (\sin(\theta 2))^2 + \gamma xy \cdot \sin(\theta 2) \cdot \cos(\theta 2)$$

$$\varepsilon C = \varepsilon x \cdot (\cos(\theta 3))^2 + \varepsilon y \cdot (\sin(\theta 3))^2 + \gamma xy \cdot \sin(\theta 3) \cdot \cos(\theta 3)$$

$$\text{Find}(\varepsilon x, \varepsilon y, \gamma xy) \rightarrow \begin{bmatrix} 40 \\ 20 \cdot \dfrac{\left(98 \cdot \sin(45 \cdot \deg)^2 - 39\right)}{\sin(45 \cdot \deg)^2} \\ 60 \cdot \dfrac{\left(32 \cdot \sin(45 \cdot \deg)^2 - 29\right)}{\sin(45 \cdot \deg)} \cdot \dfrac{\cos(45 \cdot \deg)}{\left(\sin(45 \cdot \deg)^2 - 1\right)} \end{bmatrix}$$

Insert the vectors elements to calculate the strains

$$\varepsilon x := 40 = (40)$$

$$\varepsilon y := 20 \cdot \frac{\left(98 \cdot \sin(45 \cdot \deg)^2 - 39\right)}{\sin(45 \cdot \deg)^2} = (400)$$

$$\gamma xy := 60 \cdot \frac{\left(32 \cdot \sin(45 \cdot \deg)^2 - 29\right)}{\sin(45 \cdot \deg)} \cdot \frac{\cos(45 \cdot \deg)}{\left(\sin(45 \cdot \deg)^2 - 1\right)} = (1560)$$

To solve the system of equations:
Type the word Given.
Type the equations and inequalities below the word Given.
Type the Find function entering the unknowns as arguments.
Press Ctrl+. (the control key followed by a period). Mathcad displays the right arrow.
Click outside the Find function. Mathcad places the results in a vector. The results are in the same order as the arguments to the Find function.
Once the vector is displayed copy and paste the vector elements to the variables shown on the left for strain in x and y direction

13.5 CONTACT STRESSES

The following MathCAD worksheet was designed to calculate contact stresses between two solids.

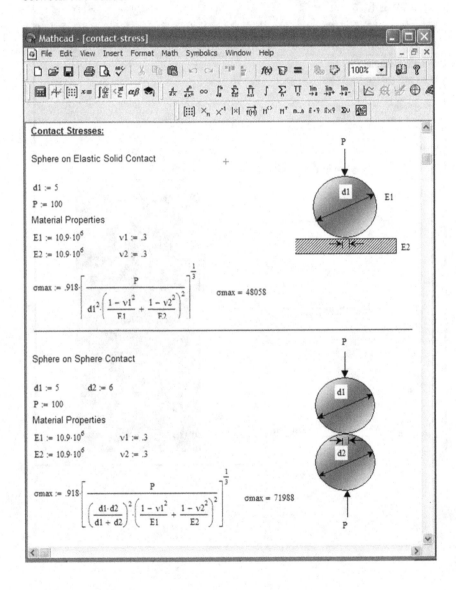

13.6 PRESSURE VESSELS

The following MathCAD worksheet was designed to calculate wall stresses on pressure vessels.

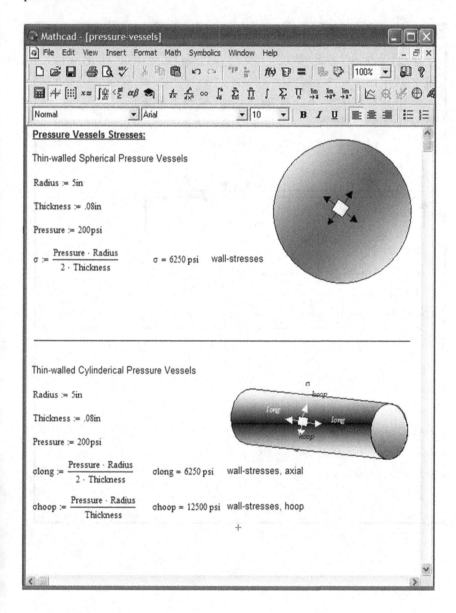

Pressure Vessels Stresses:

Thin-walled Spherical Pressure Vessels

Radius := 5in

Thickness := .08in

Pressure := 200psi

$\sigma := \dfrac{\text{Pressure} \cdot \text{Radius}}{2 \cdot \text{Thickness}}$ $\sigma = 6250\,\text{psi}$ wall-stresses

Thin-walled Cylinderical Pressure Vessels

Radius := 5in

Thickness := .08in

Pressure := 200psi

$\sigma\text{long} := \dfrac{\text{Pressure} \cdot \text{Radius}}{2 \cdot \text{Thickness}}$ $\sigma\text{long} = 6250\,\text{psi}$ wall-stresses, axial

$\sigma\text{hoop} := \dfrac{\text{Pressure} \cdot \text{Radius}}{\text{Thickness}}$ $\sigma\text{hoop} = 12500\,\text{psi}$ wall-stresses, hoop

13.7   SHEAR AND MOMENT DIAGRAMS

The following MathCAD worksheet was developed to determine the shear and moment diagram of a beam.

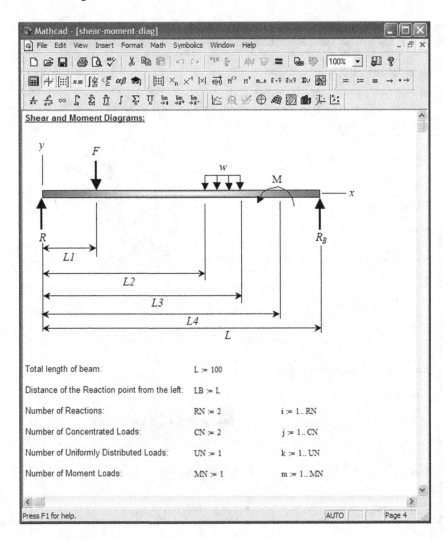

Shear and Moment Diagrams:

Total length of beam:	$L := 100$	
Distance of the Reaction point from the left:	$LB := L$	
Number of Reactions:	$RN := 2$	$i := 1..RN$
Number of Concentrated Loads:	$CN := 2$	$j := 1..CN$
Number of Uniformly Distributed Loads:	$UN := 1$	$k := 1..UN$
Number of Moment Loads:	$MN := 1$	$m := 1..MN$

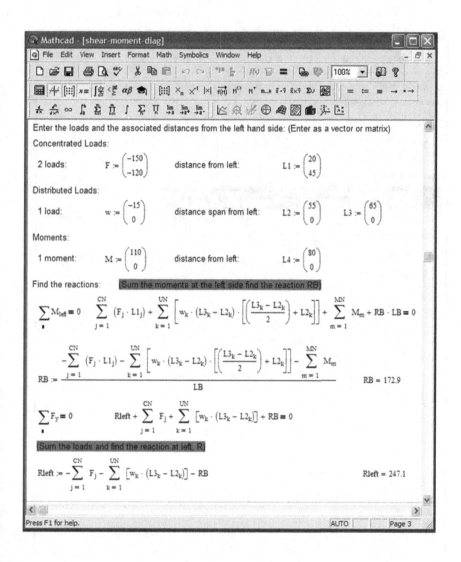

Enter the loads and the associated distances from the left hand side: (Enter as a vector or matrix)

Concentrated Loads:

2 loads: $F := \begin{pmatrix} -150 \\ -120 \end{pmatrix}$ distance from left: $L1 := \begin{pmatrix} 20 \\ 45 \end{pmatrix}$

Distributed Loads:

1 load: $w := \begin{pmatrix} -15 \\ 0 \end{pmatrix}$ distance span from left: $L2 := \begin{pmatrix} 55 \\ 0 \end{pmatrix}$ $L3 := \begin{pmatrix} 65 \\ 0 \end{pmatrix}$

Moments:

1 moment: $M := \begin{pmatrix} 110 \\ 0 \end{pmatrix}$ distance from left: $L4 := \begin{pmatrix} 80 \\ 0 \end{pmatrix}$

Find the reactions: **Sum the moments at the left side find the reaction RB**

$$\sum M_{left} = 0 \qquad \sum_{j=1}^{CN} (F_j \cdot L1_j) + \sum_{k=1}^{UN} \left[w_k \cdot (L3_k - L2_k) \cdot \left[\left(\frac{L3_k - L2_k}{2} \right) + L2_k \right] \right] + \sum_{m=1}^{MN} M_m + RB \cdot LB = 0$$

$$RB := \frac{-\sum_{j=1}^{CN} (F_j \cdot L1_j) - \sum_{k=1}^{UN} \left[w_k \cdot (L3_k - L2_k) \cdot \left[\left(\frac{L3_k - L2_k}{2} \right) + L2_k \right] \right] - \sum_{m=1}^{MN} M_m}{LB} \qquad RB = 172.9$$

$$\sum F_y = 0 \qquad Rleft + \sum_{j=1}^{CN} F_j + \sum_{k=1}^{UN} \left[w_k \cdot (L3_k - L2_k) \right] + RB = 0$$

Sum the loads and find the reaction at left, R

$$Rleft := -\sum_{j=1}^{CN} F_j - \sum_{k=1}^{UN} \left[w_k \cdot (L3_k - L2_k) \right] - RB \qquad Rleft = 247.1$$

Press F1 for help.   AUTO   Page 3

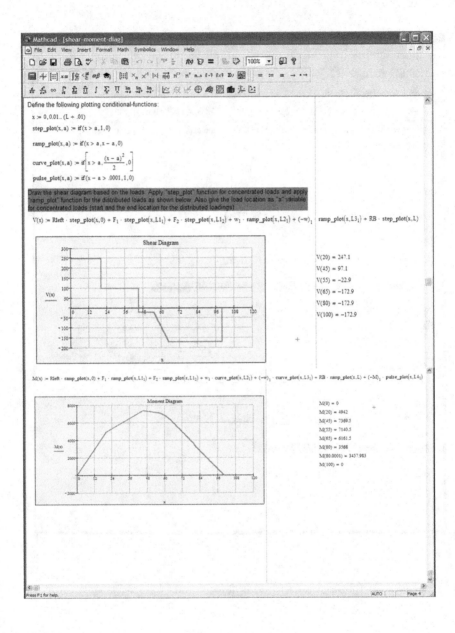

Define the following plotting conditional-functions:

x := 0, 0.01 .. (L + .01)

step_plot(x, a) := if(x > a, 1, 0)

ramp_plot(x, a) := if(x > a, x − a, 0)

curve_plot(x, a) := if$\left[x > a, \dfrac{(x − a)^2}{2}, 0 \right]$

pulse_plot(x, a) := if(x − a > .0001, 1, 0)

Draw the shear diagram based on the loads. Apply "step_plot" function for concentrated loads and apply "ramp_plot" function for the distributed loads as shown below. Also give the load location as "a" variable for concentrated loads (start and the end location for the distributed loadings)

V(x) := Rleft · step_plot(x, 0) + F$_1$ · step_plot(x, L1$_1$) + F$_2$ · step_plot(x, L1$_2$) + w$_1$ · ramp_plot(x, L2$_1$) + (−w)$_1$ · ramp_plot(x, L3$_1$) + RB · step_plot(x, L)

Shear Diagram

V(20) = 247.1
V(45) = 97.1
V(55) = −22.9
V(65) = −172.9
V(80) = −172.9
V(100) = −172.9

M(x) := Rleft · ramp_plot(x, 0) + F$_1$ · ramp_plot(x, L1$_1$) + F$_2$ · ramp_plot(x, L1$_2$) + w$_1$ · curve_plot(x, L2$_1$) + (−w)$_1$ · curve_plot(x, L3$_1$) + RB · ramp_plot(x, L) + (−M)$_1$ · pulse_plot(x, L4$_1$)

Moment Diagram

M(0) = 0
M(20) = 4942
M(45) = 7369.5
M(55) = 7140.5
M(65) = 6161.5
M(80) = 3568
M(80.0001) = 3457.983
M(100) = 0

13.8   CURVED BEAM THEORY

The following MathCAD worksheet was developed to determine the stresses on a curved beam member.

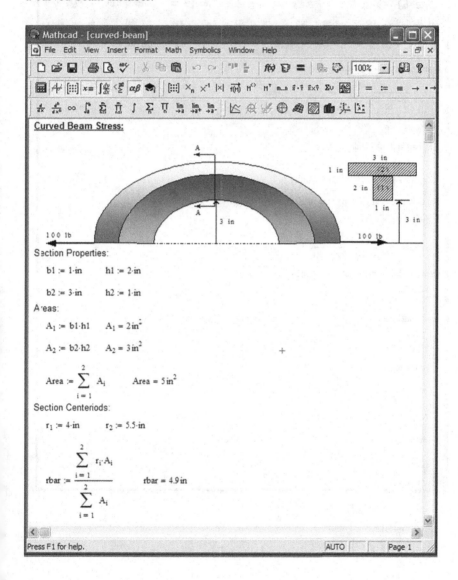

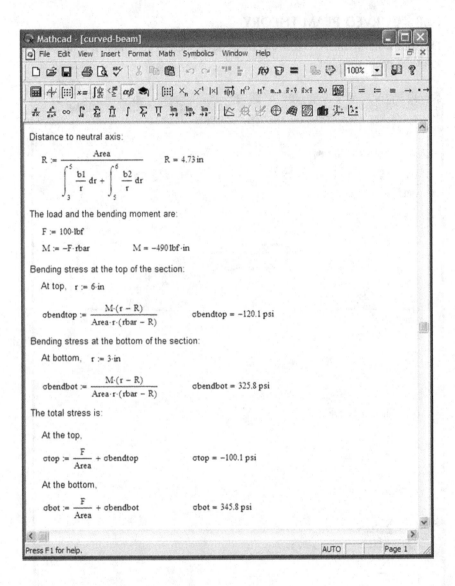

Distance to neutral axis:

$$R := \frac{Area}{\int_3^5 \frac{b1}{r}\, dr + \int_5^6 \frac{b2}{r}\, dr} \qquad R = 4.73\, in$$

The load and the bending moment are:

$F := 100 \cdot lbf$

$M := -F \cdot rbar \qquad M = -490\, lbf \cdot in$

Bending stress at the top of the section:

At top, $r := 6 \cdot in$

$$\sigma bendtop := \frac{M \cdot (r - R)}{Area \cdot r \cdot (rbar - R)} \qquad \sigma bendtop = -120.1\, psi$$

Bending stress at the bottom of the section:

At bottom, $r := 3 \cdot in$

$$\sigma bendbot := \frac{M \cdot (r - R)}{Area \cdot r \cdot (rbar - R)} \qquad \sigma bendbot = 325.8\, psi$$

The total stress is:

At the top,

$$\sigma top := \frac{F}{Area} + \sigma bendtop \qquad \sigma top = -100.1\, psi$$

At the bottom,

$$\sigma bot := \frac{F}{Area} + \sigma bendbot \qquad \sigma bot = 345.8\, psi$$

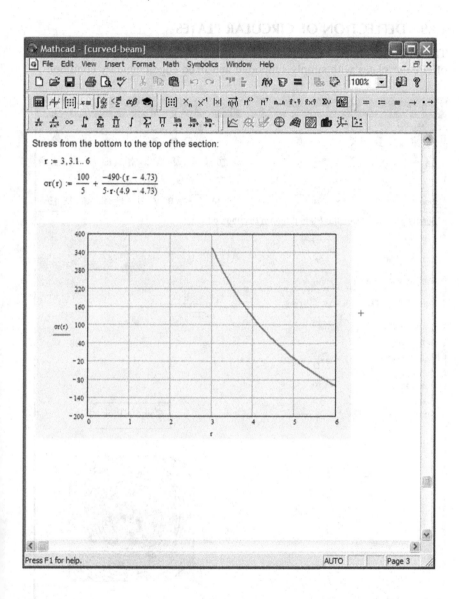

13.9 DEFLECTION OF CIRCULAR PLATES

The following MathCAD worksheets were developed to determine the deflection of circular plates with simply supported and fixed edge boundary conditions.

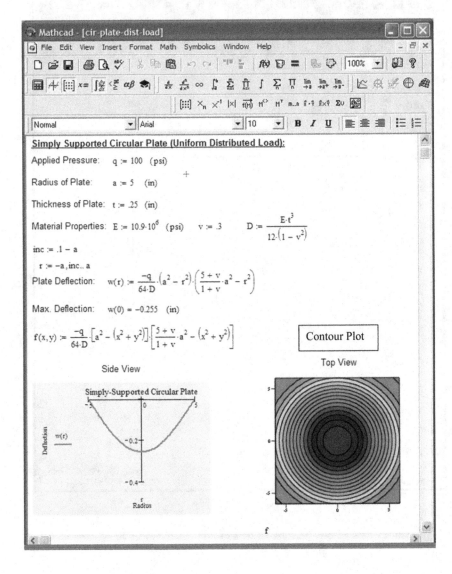

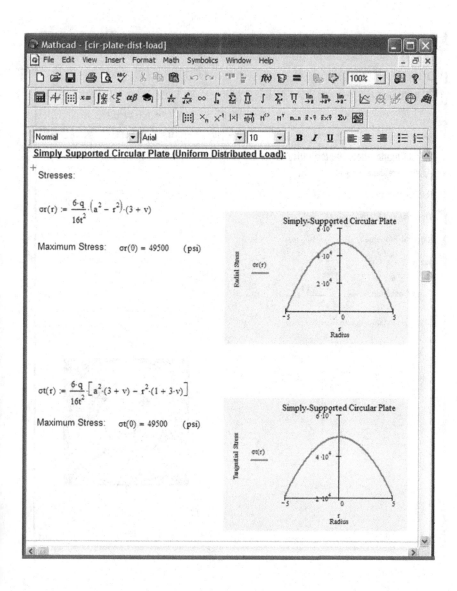

Simply Supported Circular Plate (Uniform Distributed Load):

Stresses:

$$\sigma r(r) := \frac{6 \cdot q}{16 t^2} \cdot \left(a^2 - r^2\right) \cdot (3 + v)$$

Maximum Stress: $\sigma r(0) = 49500$ (psi)

Simply-Supported Circular Plate

$$\sigma t(r) := \frac{6 \cdot q}{16 t^2} \cdot \left[a^2 \cdot (3 + v) - r^2 \cdot (1 + 3 \cdot v)\right]$$

Maximum Stress: $\sigma t(0) = 49500$ (psi)

Simply-Supported Circular Plate

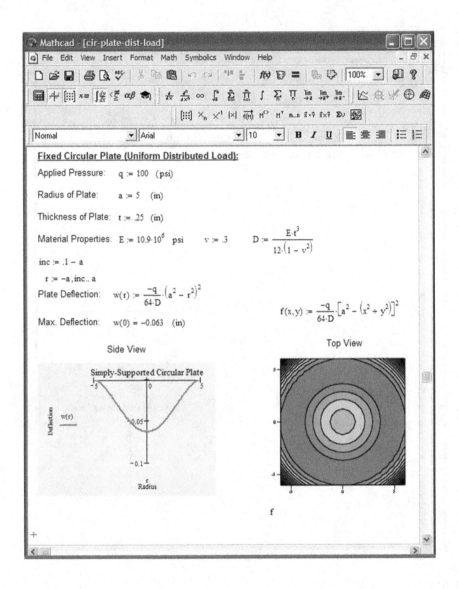

Mathcad - [cir-plate-dist-load]

File   Edit   View   Insert   Format   Math   Symbolics   Window   Help

Fixed Circular Plate (Uniform Distributed Load):

Applied Pressure: $q := 100$ (psi)

Radius of Plate: $a := 5$ (in)

Thickness of Plate: $t := .25$ (in)

Material Properties: $E := 10.9 \cdot 10^6$ psi $v := .3$ $D := \dfrac{E \cdot t^3}{12 \cdot (1 - v^2)}$

$inc := .1 - a$

$r := -a, inc .. a$

Plate Deflection: $w(r) := \dfrac{-q}{64 \cdot D} \cdot (a^2 - r^2)^2$

$f(x,y) := \dfrac{-q}{64 \cdot D} \cdot [a^2 - (x^2 + y^2)]^2$

Max. Deflection: $w(0) = -0.063$ (in)

Side View

Simply-Supported Circular Plate

Top View

f

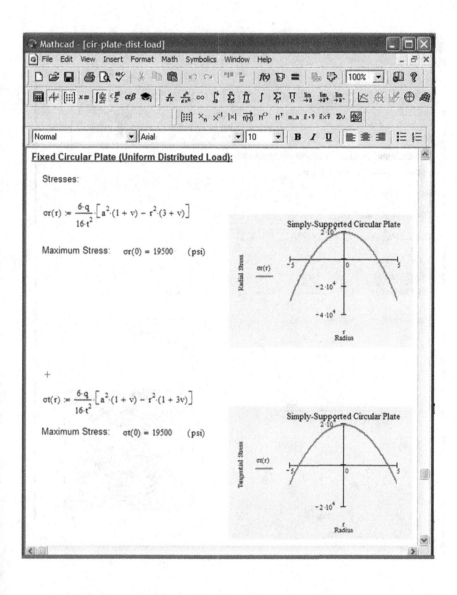

Fixed Circular Plate (Uniform Distributed Load):

Stresses:

$$\sigma r(r) := \frac{6 \cdot q}{16 \cdot t^2} \cdot \left[a^2 \cdot (1 + v) - r^2 \cdot (3 + v) \right]$$

Maximum Stress: $\sigma r(0) = 19500$ (psi)

+

$$\sigma t(r) := \frac{6 \cdot q}{16 \cdot t^2} \cdot \left[a^2 \cdot (1 + v) - r^2 \cdot (1 + 3v) \right]$$

Maximum Stress: $\sigma t(0) = 19500$ (psi)

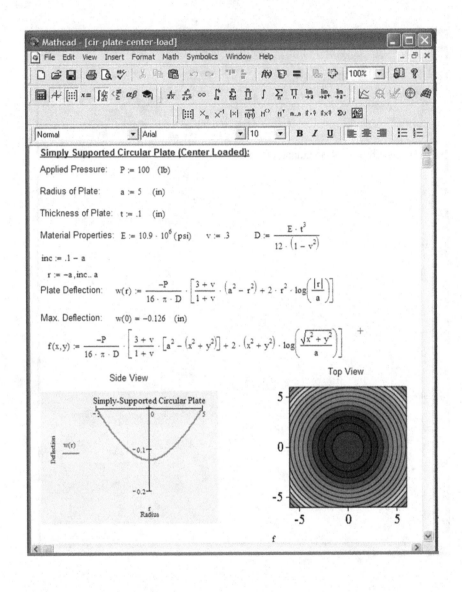

Mathcad - [cir-plate-center-load]

File   Edit   View   Insert   Format   Math   Symbolics   Window   Help

Normal   Arial   10   **B** *I* <u>U</u>

Simply Supported Circular Plate (Center Loaded):

Applied Pressure: $P := 100$ (lb)

Radius of Plate: $a := 5$ (in)

Thickness of Plate: $t := .1$ (in)

Material Properties: $E := 10.9 \cdot 10^6$ (psi) $v := .3$ $D := \dfrac{E \cdot t^3}{12 \cdot (1 - v^2)}$

$inc := .1 - a$

$r := -a, inc .. a$

Plate Deflection: $w(r) := \dfrac{-P}{16 \cdot \pi \cdot D} \cdot \left[\dfrac{3 + v}{1 + v} \cdot (a^2 - r^2) + 2 \cdot r^2 \cdot \log\left(\dfrac{|r|}{a}\right) \right]$

Max. Deflection: $w(0) = -0.126$ (in)

$f(x,y) := \dfrac{-P}{16 \cdot \pi \cdot D} \cdot \left[\dfrac{3 + v}{1 + v} \cdot \left[a^2 - (x^2 + y^2)\right] + 2 \cdot (x^2 + y^2) \cdot \log\left(\dfrac{\sqrt{x^2 + y^2}}{a}\right) \right]$

Side View   Top View

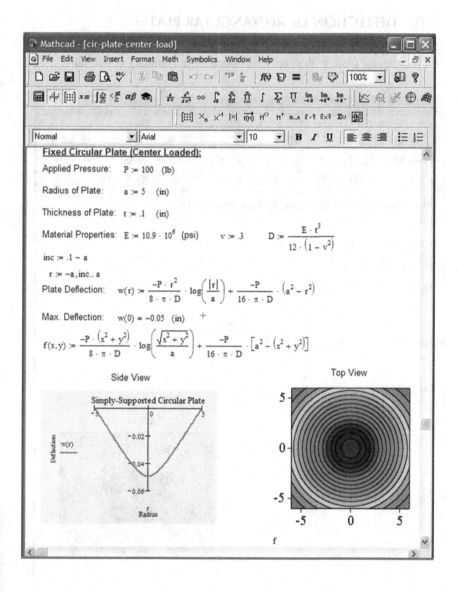

Mathcad - [cir-plate-center-load]

File Edit View Insert Format Math Symbolics Window Help

Normal   Arial   10   B I U

Fixed Circular Plate (Center Loaded):

Applied Pressure: $P := 100$ (lb)

Radius of Plate: $a := 5$ (in)

Thickness of Plate: $t := .1$ (in)

Material Properties: $E := 10.9 \cdot 10^6$ (psi) $v := .3$ $D := \dfrac{E \cdot t^3}{12 \cdot (1 - v^2)}$

$inc := .1 - a$

$r := -a, inc .. a$

Plate Deflection: $w(r) := \dfrac{-P \cdot r^2}{8 \cdot \pi \cdot D} \cdot \log\left(\dfrac{|r|}{a}\right) + \dfrac{-P}{16 \cdot \pi \cdot D} \cdot (a^2 - r^2)$

Max. Deflection: $w(0) = -0.05$ (in) $+$

$f(x, y) := \dfrac{-P \cdot (x^2 + y^2)}{8 \cdot \pi \cdot D} \cdot \log\left(\dfrac{\sqrt{x^2 + y^2}}{a}\right) + \dfrac{-P}{16 \cdot \pi \cdot D} \cdot \left[a^2 - (x^2 + y^2)\right]$

Side View   Top View

Simply-Supported Circular Plate

Deflection w(r)

Radius

f

13.10   DEFLECTION OF RECTANGULAR PLATES

The following MathCAD worksheet was developed to determine the deflection of a rectangular plate with simply supported edge boundary condition.

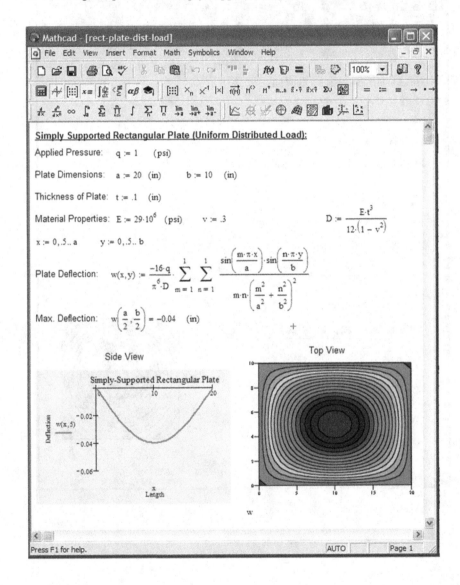

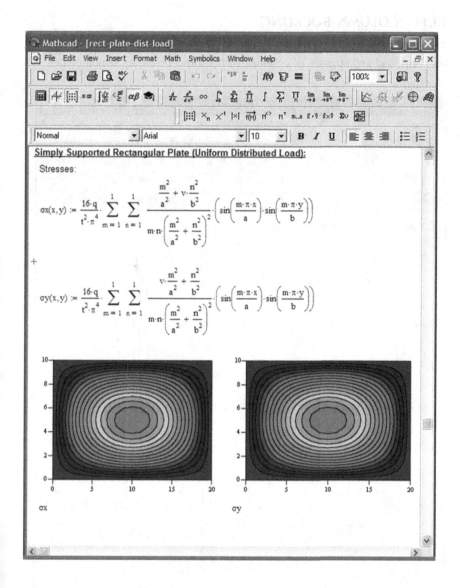

Simply Supported Rectangular Plate (Uniform Distributed Load):

Stresses:

$$\sigma x(x,y) := \frac{16 \cdot q}{t^2 \cdot \pi^4} \cdot \sum_{m=1}^{1} \sum_{n=1}^{1} \frac{\dfrac{m^2}{a^2} + v \cdot \dfrac{n^2}{b^2}}{m \cdot n \cdot \left(\dfrac{m^2}{a^2} + \dfrac{n^2}{b^2}\right)^2} \cdot \left(\sin\left(\frac{m \cdot \pi \cdot x}{a}\right) \cdot \sin\left(\frac{m \cdot \pi \cdot y}{b}\right)\right)$$

$$\sigma y(x,y) := \frac{16 \cdot q}{t^2 \cdot \pi^4} \cdot \sum_{m=1}^{1} \sum_{n=1}^{1} \frac{v \cdot \dfrac{m^2}{a^2} + \dfrac{n^2}{b^2}}{m \cdot n \cdot \left(\dfrac{m^2}{a^2} + \dfrac{n^2}{b^2}\right)^2} \cdot \left(\sin\left(\frac{m \cdot \pi \cdot x}{a}\right) \cdot \sin\left(\frac{m \cdot \pi \cdot y}{b}\right)\right)$$

σx   σy

13.11   COLUMN BUCKLING

The following MathCAD worksheets were developed to determine critical buckling load and stresses of columns with various boundary conditions.

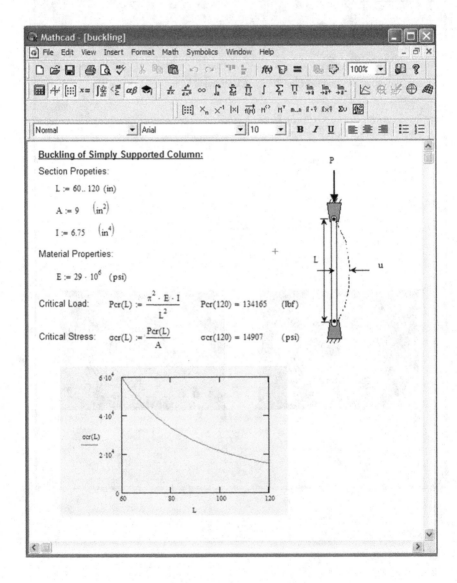

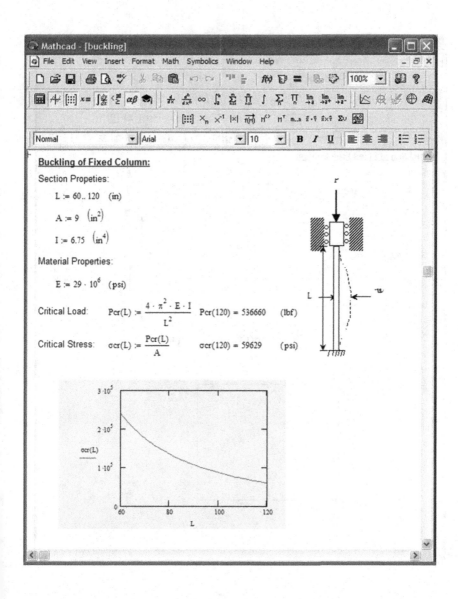

Buckling of Fixed Column:

Section Propeties:

$L := 60 .. 120$ (in)

$A := 9$ (in^2)

$I := 6.75$ (in^4)

Material Properties:

$E := 29 \cdot 10^6$ (psi)

Critical Load: $Pcr(L) := \dfrac{4 \cdot \pi^2 \cdot E \cdot I}{L^2}$ $Pcr(120) = 536660$ (lbf)

Critical Stress: $\sigma cr(L) := \dfrac{Pcr(L)}{A}$ $\sigma cr(120) = 59629$ (psi)

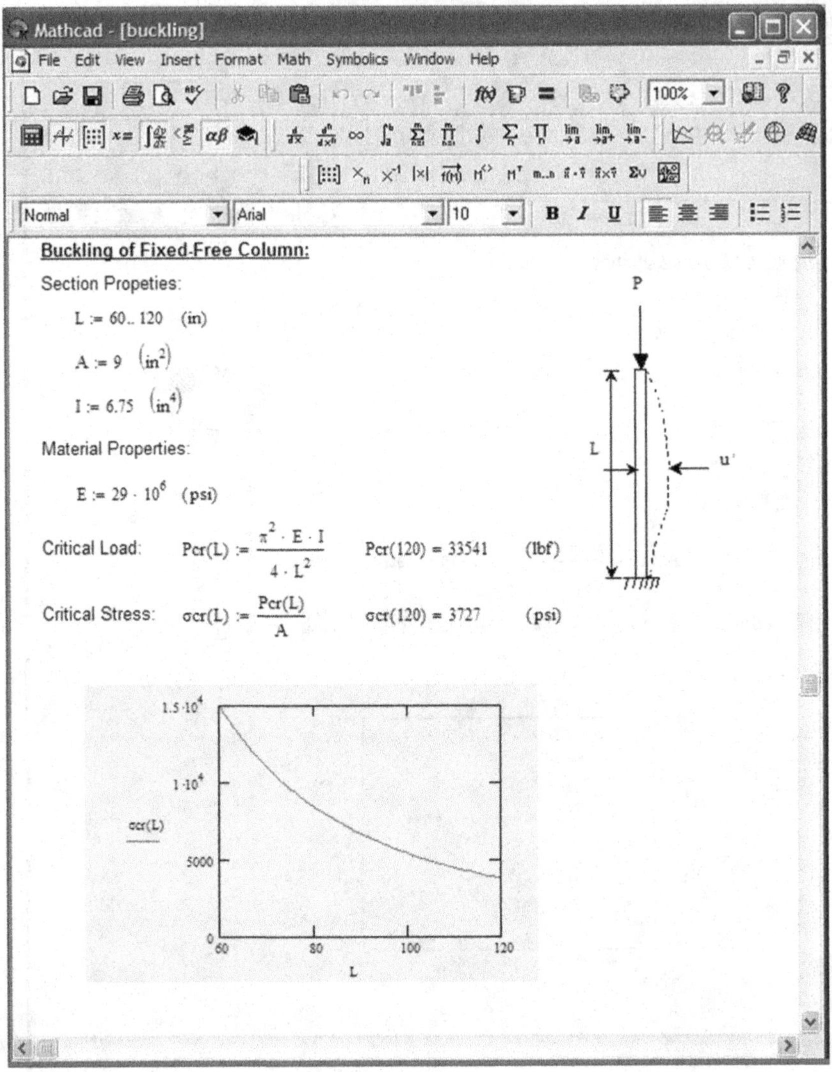
Mathcad - [buckling]
File Edit View Insert Format Math Symbolics Window Help
100%
Normal
Arial
10
Buckling of Fixed-Free Column:
Section Propeties:
L := 60.. 120 (in)
A := 9 (in^2)
I := 6.75 (in^4)
Material Properties:
E := 29 · 10^6 (psi)
Critical Load: Pcr(L) := π^2 · E · I / 4 · L^2
Pcr(120) = 33541 (lbf)
Critical Stress: σcr(L) := Pcr(L) / A
σcr(120) = 3727 (psi)
P
L
u'
1.5·10^4
1·10^4
5000
0
60
80
100
120
σcr(L)
L

13.12   PLATE BUCKLING

For plate buckling analysis, first type in the following load-coefficient data into the MathCAD worksheet, as shown below. Then, use the worksheet shown to calculate the critical load and stress of the plate.

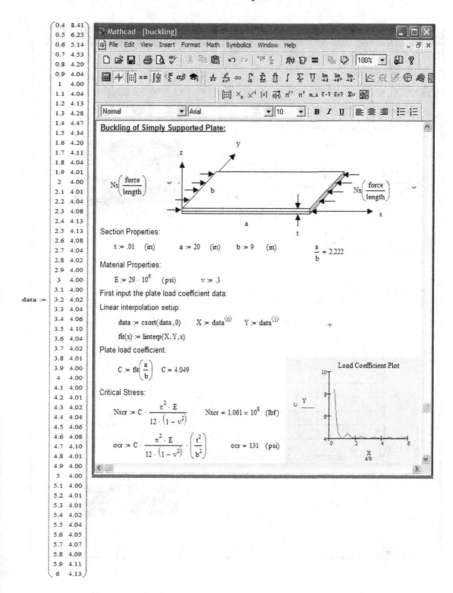

The data values shown at left:

$$data := \begin{pmatrix} 0.4 & 8.41 \\ 0.5 & 6.25 \\ 0.6 & 5.14 \\ 0.7 & 4.53 \\ 0.8 & 4.20 \\ 0.9 & 4.04 \\ 1 & 4.00 \\ 1.1 & 4.04 \\ 1.2 & 4.13 \\ 1.3 & 4.28 \\ 1.4 & 4.47 \\ 1.5 & 4.34 \\ 1.6 & 4.20 \\ 1.7 & 4.11 \\ 1.8 & 4.04 \\ 1.9 & 4.01 \\ 2 & 4.00 \\ 2.1 & 4.01 \\ 2.2 & 4.04 \\ 2.3 & 4.08 \\ 2.4 & 4.13 \\ 2.5 & 4.13 \\ 2.6 & 4.08 \\ 2.7 & 4.04 \\ 2.8 & 4.02 \\ 2.9 & 4.00 \\ 3 & 4.00 \\ 3.1 & 4.00 \\ 3.2 & 4.02 \\ 3.3 & 4.04 \\ 3.4 & 4.06 \\ 3.5 & 4.10 \\ 3.6 & 4.04 \\ 3.7 & 4.02 \\ 3.8 & 4.01 \\ 3.9 & 4.00 \\ 4 & 4.00 \\ 4.1 & 4.00 \\ 4.2 & 4.01 \\ 4.3 & 4.02 \\ 4.4 & 4.04 \\ 4.5 & 4.06 \\ 4.6 & 4.08 \\ 4.7 & 4.10 \\ 4.8 & 4.01 \\ 4.9 & 4.00 \\ 5 & 4.00 \\ 5.1 & 4.00 \\ 5.2 & 4.01 \\ 5.3 & 4.01 \\ 5.4 & 4.02 \\ 5.5 & 4.04 \\ 5.6 & 4.05 \\ 5.7 & 4.07 \\ 5.8 & 4.09 \\ 5.9 & 4.11 \\ 6 & 4.13 \end{pmatrix}$$

Buckling of Simply Supported Plate:

Section Properties:

$t := .01$ (in) $a := 20$ (in) $b := 9$ (in) $\dfrac{a}{b} = 2.222$

Material Properties:

$E := 29 \cdot 10^{6}$ (psi) $v := .3$

First input the plate load coefficient data:

Linear interpolation setup:

$data := csort(data, 0)$ $X := data^{\langle 0 \rangle}$ $Y := data^{\langle 1 \rangle}$

$fit(x) := linterp(X, Y, x)$

Plate load coefficient:

$C := fit\left(\dfrac{a}{b}\right)$ $C = 4.049$

Critical Stress:

$Nxcr := C \cdot \dfrac{\pi^{2} \cdot E}{12 \cdot (1 - v^{2})}$ $Nxcr = 1.061 \times 10^{8}$ (lbf)

$\sigma cr := C \cdot \dfrac{\pi^{2} \cdot E}{12 \cdot (1 - v^{2})} \cdot \left(\dfrac{t^{2}}{b^{2}}\right)$ $\sigma cr = 131$ (psi)

Load Coefficient Plot

13.13   TRUSS SYSTEM ANALYSIS

The following MathCAD worksheet was developed to determine the global stiffness matrix of a truss system. The truss system is the same example problem solved in chapter 10.

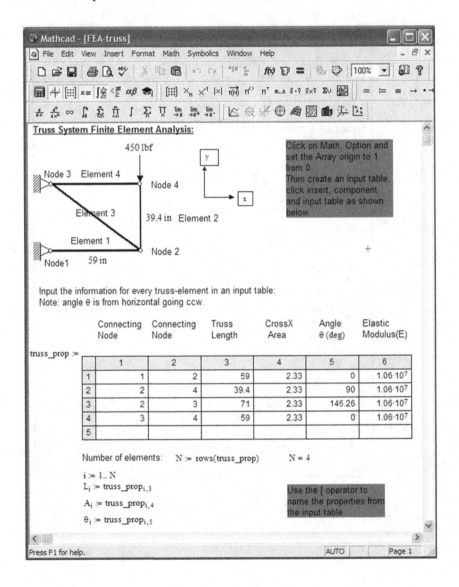

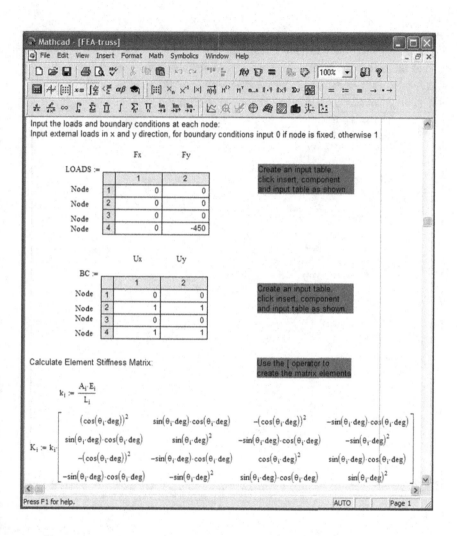

File Edit View Insert Format Math Symbolics Window Help

Input the loads and boundary conditions at each node:
Input external loads in x and y direction, for boundary conditions input 0 if node is fixed, otherwise 1

$LOADS :=$

		Fx (1)	Fy (2)
Node	1	0	0
Node	2	0	0
Node	3	0	0
Node	4	0	-450

Create an input table,
click insert, component
and input table as shown.

$BC :=$

		Ux (1)	Uy (2)
Node	1	0	0
Node	2	1	1
Node	3	0	0
Node	4	1	1

Create an input table,
click insert, component
and input table as shown.

Calculate Element Stiffness Matrix:

Use the [operator to
create the matrix elements

$$k_i := \frac{A_i \cdot E_i}{L_i}$$

$$K_i := k_i \cdot \begin{bmatrix} (\cos(\theta_i \cdot \deg))^2 & \sin(\theta_i \cdot \deg) \cdot \cos(\theta_i \cdot \deg) & -(\cos(\theta_i \cdot \deg))^2 & -\sin(\theta_i \cdot \deg) \cdot \cos(\theta_i \cdot \deg) \\ \sin(\theta_i \cdot \deg) \cdot \cos(\theta_i \cdot \deg) & \sin(\theta_i \cdot \deg)^2 & -\sin(\theta_i \cdot \deg) \cdot \cos(\theta_i \cdot \deg) & -\sin(\theta_i \cdot \deg)^2 \\ -(\cos(\theta_i \cdot \deg))^2 & -\sin(\theta_i \cdot \deg) \cdot \cos(\theta_i \cdot \deg) & \cos(\theta_i \cdot \deg)^2 & \sin(\theta_i \cdot \deg) \cdot \cos(\theta_i \cdot \deg) \\ -\sin(\theta_i \cdot \deg) \cdot \cos(\theta_i \cdot \deg) & -\sin(\theta_i \cdot \deg)^2 & \sin(\theta_i \cdot \deg) \cdot \cos(\theta_i \cdot \deg) & \sin(\theta_i \cdot \deg)^2 \end{bmatrix}$$

Press F1 for help.   AUTO   Page 1

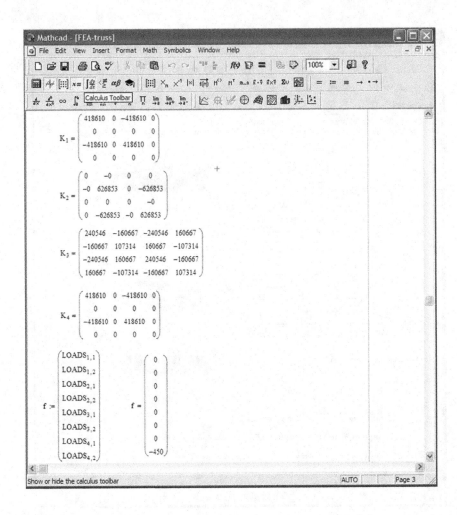

$$K_1 = \begin{pmatrix} 418610 & 0 & -418610 & 0 \\ 0 & 0 & 0 & 0 \\ -418610 & 0 & 418610 & 0 \\ 0 & 0 & 0 & 0 \end{pmatrix}$$

$$K_2 = \begin{pmatrix} 0 & -0 & 0 & 0 \\ -0 & 626853 & 0 & -626853 \\ 0 & 0 & 0 & -0 \\ 0 & -626853 & -0 & 626853 \end{pmatrix}$$

$$K_3 = \begin{pmatrix} 240546 & -160667 & -240546 & 160667 \\ -160667 & 107314 & 160667 & -107314 \\ -240546 & 160667 & 240546 & -160667 \\ 160667 & -107314 & -160667 & 107314 \end{pmatrix}$$

$$K_4 = \begin{pmatrix} 418610 & 0 & -418610 & 0 \\ 0 & 0 & 0 & 0 \\ -418610 & 0 & 418610 & 0 \\ 0 & 0 & 0 & 0 \end{pmatrix}$$

$$f := \begin{pmatrix} LOADS_{1,1} \\ LOADS_{1,2} \\ LOADS_{2,1} \\ LOADS_{2,2} \\ LOADS_{3,1} \\ LOADS_{3,2} \\ LOADS_{4,1} \\ LOADS_{4,2} \end{pmatrix} \qquad f = \begin{pmatrix} 0 \\ 0 \\ 0 \\ 0 \\ 0 \\ 0 \\ 0 \\ -450 \end{pmatrix}$$

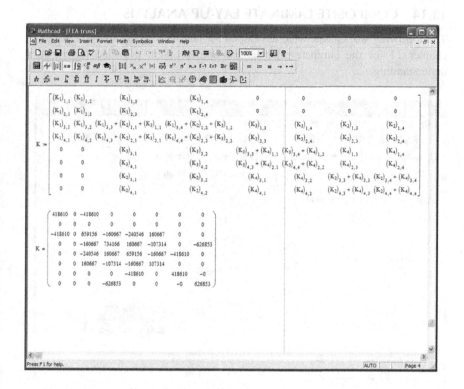

13.14   COMPOSITE LAMINATE LAY-UP ANALYSIS

The following worksheets were developed for laminate lay-up analysis. Refer to chapter 11 for the laminate analysis example problem provided for a better understanding.

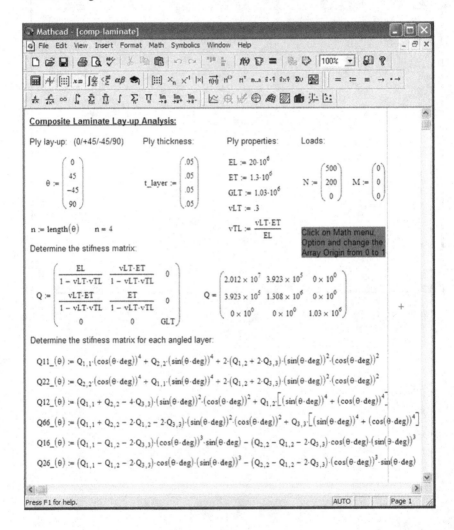

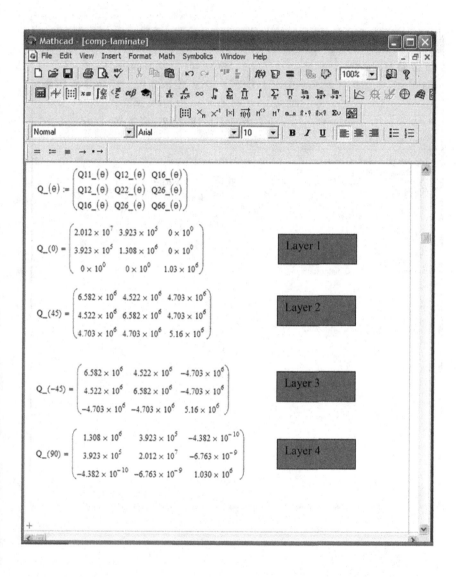

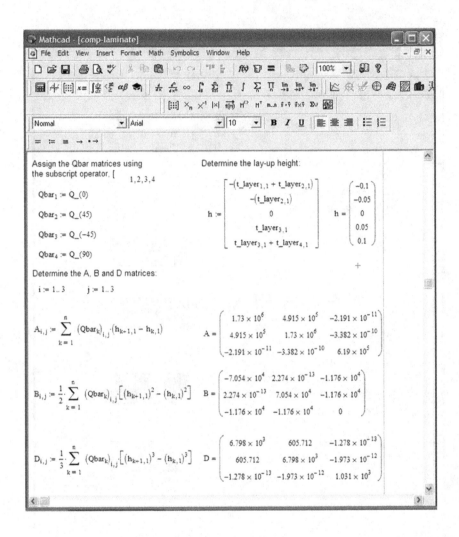

Assign the Qbar matrices using the subscript operator. [1,2,3,4

$$Qbar_1 := Q\_(0)$$

$$Qbar_2 := Q\_(45)$$

$$Qbar_3 := Q\_(-45)$$

$$Qbar_4 := Q\_(90)$$

Determine the lay-up height:

$$h := \begin{bmatrix} -(t\_layer_{1,1} + t\_layer_{2,1}) \\ -(t\_layer_{2,1}) \\ 0 \\ t\_layer_{3,1} \\ t\_layer_{3,1} + t\_layer_{4,1} \end{bmatrix} \qquad h = \begin{pmatrix} -0.1 \\ -0.05 \\ 0 \\ 0.05 \\ 0.1 \end{pmatrix}$$

Determine the A, B and D matrices:

$$i := 1..3 \qquad j := 1..3$$

$$A_{i,j} := \sum_{k=1}^{n} (Qbar_k)_{i,j} \cdot (h_{k-1,1} - h_{k,1})$$

$$A = \begin{pmatrix} 1.73 \times 10^6 & 4.915 \times 10^5 & -2.191 \times 10^{-11} \\ 4.915 \times 10^5 & 1.73 \times 10^6 & -3.382 \times 10^{-10} \\ -2.191 \times 10^{-11} & -3.382 \times 10^{-10} & 6.19 \times 10^5 \end{pmatrix}$$

$$B_{i,j} := \frac{1}{2} \cdot \sum_{k=1}^{n} (Qbar_k)_{i,j} \cdot \left[(h_{k+1,1})^2 - (h_{k,1})^2 \right]$$

$$B = \begin{pmatrix} -7.054 \times 10^4 & 2.274 \times 10^{-13} & -1.176 \times 10^4 \\ 2.274 \times 10^{-13} & 7.054 \times 10^4 & -1.176 \times 10^4 \\ -1.176 \times 10^4 & -1.176 \times 10^4 & 0 \end{pmatrix}$$

$$D_{i,j} := \frac{1}{3} \cdot \sum_{k=1}^{n} (Qbar_k)_{i,j} \cdot \left[(h_{k+1,1})^3 - (h_{k,1})^3 \right]$$

$$D = \begin{pmatrix} 6.798 \times 10^3 & 605.712 & -1.278 \times 10^{-13} \\ 605.712 & 6.798 \times 10^3 & -1.973 \times 10^{-12} \\ -1.278 \times 10^{-13} & -1.973 \times 10^{-12} & 1.031 \times 10^3 \end{pmatrix}$$

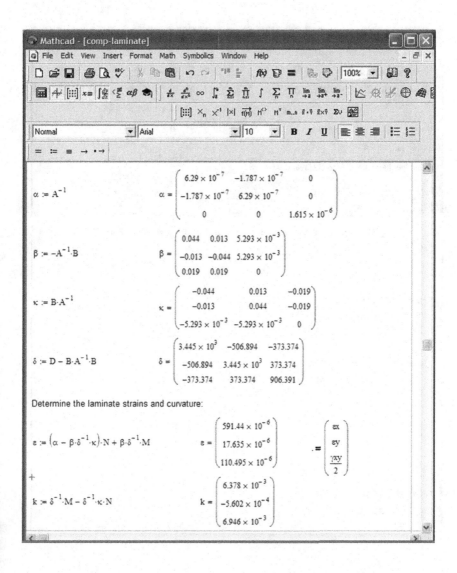

$\alpha := A^{-1}$

$$\alpha = \begin{pmatrix} 6.29 \times 10^{-7} & -1.787 \times 10^{-7} & 0 \\ -1.787 \times 10^{-7} & 6.29 \times 10^{-7} & 0 \\ 0 & 0 & 1.615 \times 10^{-6} \end{pmatrix}$$

$\beta := -A^{-1} \cdot B$

$$\beta = \begin{pmatrix} 0.044 & 0.013 & 5.293 \times 10^{-3} \\ -0.013 & -0.044 & 5.293 \times 10^{-3} \\ 0.019 & 0.019 & 0 \end{pmatrix}$$

$\kappa := B \cdot A^{-1}$

$$\kappa = \begin{pmatrix} -0.044 & 0.013 & -0.019 \\ -0.013 & 0.044 & -0.019 \\ -5.293 \times 10^{-3} & -5.293 \times 10^{-3} & 0 \end{pmatrix}$$

$\delta := D - B \cdot A^{-1} \cdot B$

$$\delta = \begin{pmatrix} 3.445 \times 10^{3} & -506.894 & -373.374 \\ -506.894 & 3.445 \times 10^{3} & 373.374 \\ -373.374 & 373.374 & 906.391 \end{pmatrix}$$

Determine the laminate strains and curvature:

$\varepsilon := \left(\alpha - \beta \cdot \delta^{-1} \cdot \kappa \right) \cdot N + \beta \cdot \delta^{-1} \cdot M$

$$\varepsilon = \begin{pmatrix} 591.44 \times 10^{-6} \\ 17.635 \times 10^{-6} \\ 110.495 \times 10^{-6} \end{pmatrix} \qquad . = \begin{pmatrix} \varepsilon x \\ \varepsilon y \\ \dfrac{\gamma xy}{2} \end{pmatrix}$$

+

$k := \delta^{-1} \cdot M - \delta^{-1} \cdot \kappa \cdot N$

$$k = \begin{pmatrix} 6.378 \times 10^{-3} \\ -5.602 \times 10^{-4} \\ 6.946 \times 10^{-3} \end{pmatrix}$$

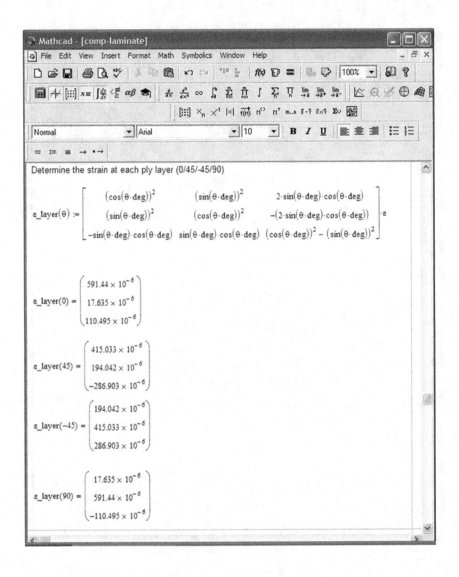

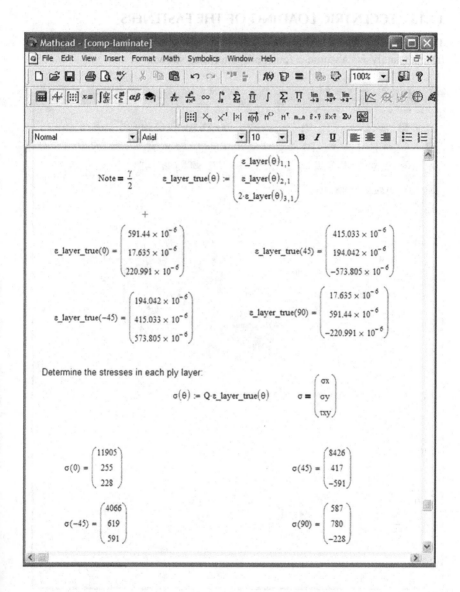

Mathcad - [comp-laminate]

File Edit View Insert Format Math Symbolics Window Help

Normal | Arial | 10 | B *I* U

$$\text{Note} = \frac{\gamma}{2} \qquad \varepsilon\_\text{layer\_true}(\theta) := \begin{pmatrix} \varepsilon\_\text{layer}(\theta)_{1,1} \\ \varepsilon\_\text{layer}(\theta)_{2,1} \\ 2\cdot\varepsilon\_\text{layer}(\theta)_{3,1} \end{pmatrix}$$

$$\varepsilon\_\text{layer\_true}(0) = \begin{pmatrix} 591.44 \times 10^{-6} \\ 17.635 \times 10^{-6} \\ 220.991 \times 10^{-6} \end{pmatrix} \qquad \varepsilon\_\text{layer\_true}(45) = \begin{pmatrix} 415.033 \times 10^{-6} \\ 194.042 \times 10^{-6} \\ -573.805 \times 10^{-6} \end{pmatrix}$$

$$\varepsilon\_\text{layer\_true}(-45) = \begin{pmatrix} 194.042 \times 10^{-6} \\ 415.033 \times 10^{-6} \\ 573.805 \times 10^{-6} \end{pmatrix} \qquad \varepsilon\_\text{layer\_true}(90) = \begin{pmatrix} 17.635 \times 10^{-6} \\ 591.44 \times 10^{-6} \\ -220.991 \times 10^{-6} \end{pmatrix}$$

Determine the stresses in each ply layer:

$$\sigma(\theta) := Q\cdot\varepsilon\_\text{layer\_true}(\theta) \qquad \sigma = \begin{pmatrix} \sigma x \\ \sigma y \\ \tau x y \end{pmatrix}$$

$$\sigma(0) = \begin{pmatrix} 11905 \\ 255 \\ 228 \end{pmatrix} \qquad \sigma(45) = \begin{pmatrix} 8426 \\ 417 \\ -591 \end{pmatrix}$$

$$\sigma(-45) = \begin{pmatrix} 4066 \\ 619 \\ 591 \end{pmatrix} \qquad \sigma(90) = \begin{pmatrix} 587 \\ 780 \\ -228 \end{pmatrix}$$

13.15   ECCENTRIC LOADING OF THE FASTENERS

The following MathCAD worksheets were developed for eccentric loading
fastener analysis. Assume a four-bolt pattern, as shown.

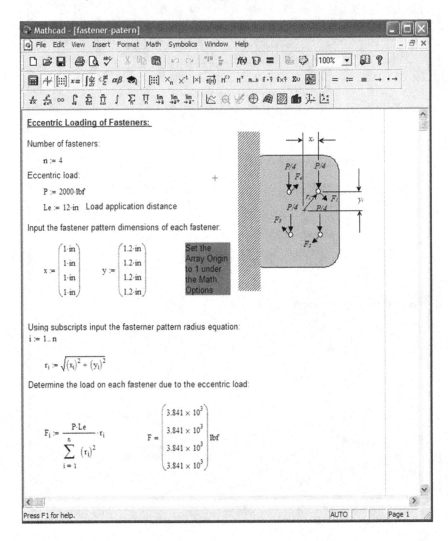

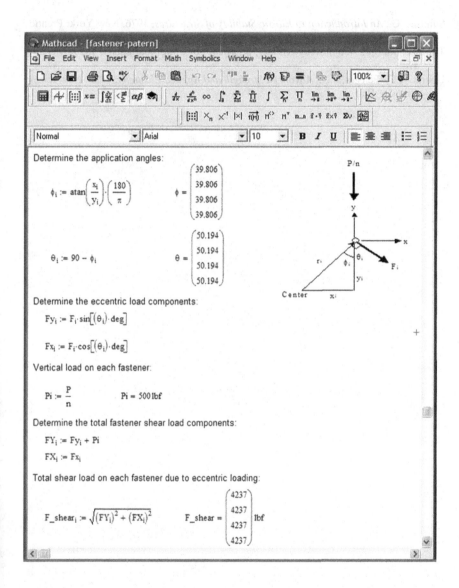

Determine the application angles:

$$\phi_i := atan\left(\frac{x_i}{y_i}\right)\cdot\left(\frac{180}{\pi}\right) \qquad \phi = \begin{pmatrix} 39.806 \\ 39.806 \\ 39.806 \\ 39.806 \end{pmatrix}$$

$$\theta_i := 90 - \phi_i \qquad \theta = \begin{pmatrix} 50.194 \\ 50.194 \\ 50.194 \\ 50.194 \end{pmatrix}$$

Determine the eccentric load components:

$$Fy_i := F_i\cdot sin\left[(\theta_i)\cdot deg\right]$$

$$Fx_i := F_i\cdot cos\left[(\theta_i)\cdot deg\right]$$

Vertical load on each fastener:

$$Pi := \frac{P}{n} \qquad Pi = 500\,lbf$$

Determine the total fastener shear load components:

$$FY_i := Fy_i + Pi$$

$$FX_i := Fx_i$$

Total shear load on each fastener due to eccentric loading:

$$F\_shear_i := \sqrt{(FY_i)^2 + (FX_i)^2} \qquad F\_shear = \begin{pmatrix} 4237 \\ 4237 \\ 4237 \\ 4237 \end{pmatrix} lbf$$

REFERENCES

Agarwal, B. D., Broutman, L.J., *Analysis and Performance of Fiber Composites*, 1990. New York: John Wiley & Sons.

Beer, F.P., Johnston, E.R., DeWolf, J.T., *Mechanics of Material*, 2002. New York: McGraw Hill.

Gurdal, Z., Haftka, R.T., Hajela, P., *Design and Optimization of Laminated Composite Materials*. 1999, New York: John Wiley & Sons.

Logan, D.L., *Finite Element Methods*, 2002. California: Wadsworth.

Simitses, G., *An Introduction to Elastic Stability of Structures*, 1976. New York: Prentice Hall.

Timoshenko, S.P., Woinowsky-Krieger, S., *Theory of Plates and Shells*, 1959. McGraw Hill.

Ugural, A.C., Fenster, S.K., *Advanced Strength and Applied Elasticity*, 1995. New Jersey: Prentice Hall.

Index